Winds of Change

WINDS OF CHANGE

The Environmental Movement and the Global Development of the Wind Energy Industry

Ion Bogdan Vasi

OXFORD
UNIVERSITY PRESS

OXFORD

UNIVERSITY PRESS

Oxford University Press, Inc., publishes works that further
Oxford University's objective of excellence
in research, scholarship, and education.

Oxford   New York
Auckland   Cape Town   Dar es Salaam   Hong Kong   Karachi
Kuala Lumpur   Madrid   Melbourne   Mexico City   Nairobi
New Delhi   Shanghai   Taipei   Toronto

With offices in
Argentina   Austria   Brazil   Chile   Czech Republic   France   Greece
Guatemala   Hungary   Italy   Japan   Poland   Portugal   Singapore
South Korea   Switzerland   Thailand   Turkey   Ukraine   Vietnam

Copyright © 2011 by Oxford University Press, Inc.

Published by Oxford University Press, Inc.
198 Madison Avenue, New York, New York 10016

www.oup.com

Oxford is a registered trademark of Oxford University Press

Library of Congress Cataloging-in-Publication Data
Vasi, Ion Bogdan.
Winds of change: the environmental movement and the global
development of the wind energy industry / Ion Bogdan Vasi.
    p. cm.
Includes bibliographical references and index.
ISBN 978-0-19-974692-7
1. Wind power industry.   2. Environmentalism.   I. Title.
HD9502.5.W55V37 2011
333.9'2—dc22   2010016162

9 8 7 6 5 4 3 2 1

Printed in the United States of America
on acid-free paper

Contents

Preface and Acknowledgments

This book has two origins. One goes back to the mid-1990s, when I first encountered wind turbines. I was riding a bicycle through northern Germany, on a road that wound through a rural area dotted with large but sleek turbines. I was immediately impressed by their elegance and technological sophistication. Standing right next to a machine that produced electricity for almost a hundred houses, I was able to have a conversation with my guide in a normal voice. There was no air pollution, no radiation, and the fuel was free. Many local residents owned shares in the wind farm and were very proud of the turbines' presence in their "backyard." Later I remembered a song by Ella Fitzgerald and I thought, This Could be the Start of Something Big!

The other origin of this book goes back to the second-half of the 1990s, when more and more people—including me—became concerned about global climate change. The Kyoto Protocol was adopted in 1997 and, at the end of the twentieth century, a number of states and local governments were already reducing their emissions of greenhouse gases. My doctoral education at Cornell University exposed me to the most innovative theories about social movements and collective action, and led me to study what I termed the ultimate social dilemma: global climate change. I wrote my dissertation on local actions against global climate change, seeking to understand why some communities were taking steps to reduce their carbon footprint while others were not. During my research, I became more and more interested in renewable energy and, in particular, wind power.

I decided to write a book about wind energy at about the same time I decided to run my first marathon, in 2005. Little did I know that running a marathon and writing a book have so much in common! For example, a certain amount of pain and anxiety is involved in both experiences. While I did not suffer physically when I conducted research and typed the manuscript, I sometimes felt that my writing was painfully slow, and I asked myself frequently, "Will I be able to finish what I've started when I want to?" Also, euphoric moments punctuate the monotony of both experiences. Writing a good section thrilled me as much as finishing a good training run. Finally, both experiences are simultaneously solitary and social. The solitude of running is famously captured by the title of Alan Sillitoe's story *The Loneliness of the Long Distance Runner*, and the solitude of writing is captured by the title of Wright Morris's collection of novels *The Loneliness of the Long Distance Writer*. But, in both writing and running, a good outcome depends on the amount of social support received—in fact, the hero in Sillitoe's story does not finish the run because he does not have the necessary support. If I was able to finish the book—and eight marathons—by June 2010, it is mainly because I had help.

During the race to finish this book, and even before I began, I was fortunate to have a great deal of support from colleagues, students, friends, and loved ones. Sidney Tarrow deserves most thanks for constantly advising me, beginning with my dissertation study and continuing through the research and writing of this book. David Strang, Michael Lounsbury, and Michael Macy also deserve special thanks for offering ideas on literatures and helping me improve my understanding of social change processes. I owe thanks to Daniel Sherman and Felix Kolb for offering extensive comments on how to structure the book. I also owe a lot to Chuck Tilly, who was extremely generous and commented on an early draft of the manuscript even though he was not physically well. Braydon King, Doug McAdam, David Meyer, Hayagreeva Rao, Jackie Smith, Sarah Soule, and Mayer Zald also deserve thanks for offering suggestions and encouragement.

Along the way, I was very fortunate to have a number of students from Columbia University who helped me with the research; without their help, this book would have taken significantly longer to complete. I owe thanks to Eliav Bitan, Whitney Blake, William Covin, Gabriel Cowles, Molly DeSalle, Myriame Gabay, Elyse Hottel, Kara Kirchhoff, Jessica McHugh, Cathleen Monahan, Andrew Miller, Carlos Rymer, and Alla Sobel. I also owe thanks to my good friend Sarah Coleman, who read each chapter and helped me edit the manuscript. James Cook, the patient and creative editor from Oxford University Press, and three anonymous reviewers provided essential advice for refining my arguments: I thank all of them.

The people who agreed to be interviewed and to provide various documents made this book possible; they are too many to list. But it is obvious that I couldn't have completed this project without their help. I couldn't think of a better way to thank them than to donate all the money I will receive from writing this book to a company that builds new wind farms.

Finally, I could not have completed this book without support from my family. I always knew I could count on my parents for moral support. My wife, Mihaela, has been as patient and supportive as I could have ever wished. And thinking about our wonderful daughters, Anna and Iris, gave me the strength I needed to complete what I started. I dedicate this book, with much love and gratitude, to them.

Winds of Change

Introduction

The Wind Energy Industry and the Environmental Movement

Wind energy conversion is a fascinating field to study, if only because its past has been so checkered and its exact future is so uncertain. Unlike the aerospace industry, the computer industry, and almost any other successful industry you can name, wind energy…has been around for thousands of years. It's a technology that has been reinvented numerous times. We are left with the promise and the drive to succeed despite daunting obstacles.
—Darrell Dodge, "An Illustrated History of Wind Power Development," http://www.telosnet.com/wind/20th.html (accessed April 2007)

The Puzzling Development of the Wind Energy Industry

The answer to future problems of energy supply seemed to be blowing in the wind in 1980. The global generating capacity of wind power was only 10 megawatts (MW)—enough electricity to power approximately three thousand "average" U.S. homes, but many policymakers and energy analysts were optimistic about the future of the wind energy industry. That year the U.S. House of Representatives had voted "to harness the wind and put it to work for a brighter America" by passing the Wind Energy Systems Act, a bill that aimed to reach a total capacity from wind energy systems of at least 800 MW by 1988.[1] In the same year, analysts from the U.S. Department of Energy were estimating that the United States would likely get 20 percent of its electricity from wind by the year 2000.[2]

Indeed, as new materials and technologies became available, a new era of wind energy began in the 1980s. By 1985, the United States had an installed capacity of more than 1 gigawatt (GW) from wind energy, far exceeding the goal of 800 MW by 1988, as set forth in the Wind Energy Systems Act. Five years

3

later the worldwide capacity had almost doubled, and almost three-quarters of the installed capacity in 1990 was in the United States. Other countries had also installed significant wind power capacity; for example, by 1991 Denmark had installed over 400 MW, while Germany had installed over 100 MW.

The rapid rate of growth continued through the 1990s, and by the beginning of the twenty-first century the global installed capacity had reached more than 18 GW (18,000 MW). The wind power industry's annual growth rate had averaged around 30 percent since the mid-1990s and reached a record of 42 percent in 2005, making it the world's fastest-growing energy industry and one of the fastest-growing business sectors. By the beginning of 2009, the installed capacity from wind energy was almost 121 GW (121,000 MW) (see figure 1). As some energy analysts noted, "The market [for wind energy] has exploded."[3] Moreover, the global wind market is expected to continue to grow at such a fast rate that it could reach over 240 GW of total installed capacity by the year 2012.[4]

Despite this impressive growth, however, the wind energy industry has not reached its true potential. Wind energy accounted for about 1 percent of the electricity generated in the United States in 2008, significantly less than was projected by the Department of Energy analysts more than a quarter century ago. This is particularly disappointing because the United States has some of the best wind energy resources in the world. For instance, in 1991 a national wind resource inventory taken by the U.S. Department of Energy reported that three states—North Dakota, Kansas, and Texas—had enough viable wind energy to satisfy the electricity needs of the entire nation.[5] Similarly, in the United Kingdom wind energy accounted for about 1.5 percent of the electricity generated in 2008, although the country has enough onshore and offshore wind potential to power itself three times over.[6] Globally, the wind energy industry accounted for approximately 1.5 percent of electricity used at the end of 2008, even though the world's winds could theoretically supply more than thirty times the current worldwide electricity demand.[7]

Not only is worldwide wind energy production far from reaching its vast potential, it is also highly uneven. While some countries have built multiple wind farms, many countries have little or no wind power on a commercial scale. As figure 2 shows, ten countries accounted for almost 85 percent of the global wind energy installed capacity at the end of 2008.

What drives the relatively fast but uneven growth of the wind energy industry worldwide? In other words, why is it that wind power stands out as one of the splashiest success stories in renewable energy, but has failed to reach its full potential and developed irregularly in different parts of the world? This uneven global development is puzzling. Many countries and regions with some of the best wind energy potential do not have the highest installed capacity, as is the case with the United Kingdom, which has by far the best wind energy potential of all the European Union countries. Recent studies show the United Kingdom has the strongest, most dependable, and most convenient onshore winds, as well as the highest offshore wind

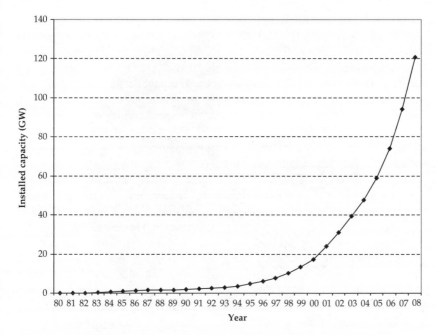

Figure I.1. Global cumulative installed capacity (MW) from wind energy, 1980–2008. World Watch Institute; European Wind Energy Association; Global Wind Energy Council.

potential.[8] However, the United Kingdom ranked sixth in Europe in installed wind power capacity at the beginning of 2008 and was generating almost 10 percent less wind energy than Germany. Conversely, some leading U.S. regions and states in installed wind power capacity do not have the best wind power resources. For example, California was the leading U.S. wind energy producer in 2005 although it was ranked just seventeenth in wind generation potential. North Dakota has been frequently called the "Saudi Arabia of wind"—the state's persistent winds are capable of generating enough power for more than a quarter of the nation. Yet in 2006 it ranked fifteenth in installed capacity.[9]

Two perspectives dominate accounts of the growth of the wind energy industry—and research on new industries generally. The technological perspective argues that the development of new industries is influenced primarily by technological innovations and traditions. In this view, the global growth of the wind energy industry results from decreasing costs in wind power generation due to continuous improvements in blade, gearbox, and generator technology, while cross-national variation results from differences in technological approaches. Numerous academic studies have used this framework to show that differences in technological approaches account for the early success of

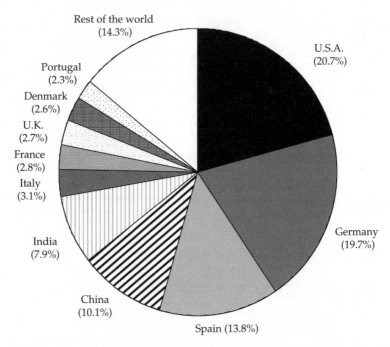

Figure I.2. Top ten countries in terms of wind power installed capacity by the end of 2008—percentage of total global capacity. Adapted from World Wind Energy Association, "World Wind Energy Report 2008."

wind turbine manufacturers in some countries and not others (Heymann 1998; Est 1999; Karnøe, Kristensen, and Andersen 1999; Johnson and Jacobsson 2000; Garud and Karnøe 2003; Boon 2008). Wind power advocates frequently employ this framework and argue that the industry's global development is inevitable, given that it is driven by continuous technological advances. Paul Gipe—a well-known wind energy pioneer—wrote in the mid-1990s, "From the deserts of California to the shores of the North Sea, wind energy has come of age as a commercial generating technology. Wind turbines now provide commercial bulk power in California, Hawaii, Minnesota, Alberta, Denmark, Germany, Great Britain, Spain, Italy, Greece, the Netherlands, India and China, and the list continues to grow" (Gipe 1995).

The market perspective, on the other hand, argues that the emergence of new industries is shaped by various economic forces. Academic research taking this perspective examines how the interaction between factors such as the adoption of specific energy policies, the supply of wind turbine components, or the deregulation of electric utilities determines wind energy price and market penetration. Several studies have compared public policies such as power purchase agreements, investment and production incentives, and renewable set-asides to assess whether they create efficient markets for renewable energy (Redlinger, Andersen, and Morthorst 2002; Lauber 2005). Energy professionals often debate the growth of the renewable energy sector in terms of the costs associated with specific policies. A 2008 report of the International Energy Agency exemplifies this approach:

> The group of countries with the highest effectiveness...used feed-in tariffs (FITs) to encourage wind power deployment. Their success in deploying onshore wind stems from high investment stability guaranteed by the long term FITs, an appropriate framework with low administrative and regulatory barriers, and relatively favorable grid access conditions. In 2005, the average remuneration levels in these countries (USD 0.09–0.11/kWh [kilowatt-hours]) were lower than those in countries applying quota obligation systems with tradable green certificates. (TGCs) (USD 0.13–0.17/kWh)[10]

These two dominant perspectives cannot fully explain the puzzling development of the wind energy industry. The first perspective can explain why Denmark, which adopted a specific approach to technology, was successful in wind turbine innovation; it cannot explain why the wind energy industry also developed in countries and regions that did not use the Danish technological approach. The second perspective can explain why the wind energy industry developed faster in countries and regions that adopted policies such as feed-in tariffs, but cannot explain why only certain countries and regions adopted these policies.

An uncommon approach to understanding burgeoning industries considers the influence of social movements. This perspective acknowledges the importance of technological or economic factors, but centers primarily on the role played by social movements in the growth of new industries. In the case of the wind energy industry, this perspective analyzes how the environmental movement has influenced the industry's growth in various national and subnational contexts. This book argues that the global development of the wind energy industry cannot be understood without examining the interactions of environmental activists and organizations with governments, energy-sector actors, various institutions, and the general public over the last four decades.

Standard Perspectives on Industry Creation and Wind Energy

Much research on the wind energy industry and other new industries employs a technological perspective. Studies such as those by Gipe (1995); Ackermann (2005); Bhadra, Kastha, and Banerjee (2005); and Heier (2006) attribute the growth of the wind energy industry to improved technology and reductions in the cost of electricity produced from wind, and not to the social and environmental implications of wind power technology. The following comment, made by renewable energy expert Werner Zittel, summarizes this approach:

> We have seen more than ten years of unprecedented growth in this sector driven by a wide variety of factors ranging from cost reductions to access to new high-wind resources and better grid regulations.... This is *not about morals or environment* but the *commercial reality* that wind, coupled with hydro, solar, biomass and geothermal energy is not only a rapid and cost effective alternative but one that could deliver all our energy requirements within the first half of this century.[11]

Engineers and other energy specialists frequently employ a technological perspective when discussing the feasibility of wind power. David MacKay argues in his acclaimed book *Sustainable Energy—Without the Hot Air* that the future of the wind energy industry in Britain depends on solving two technical problems:

> We need to solve two problems—lulls (long periods with small renewable production), and slews (short-term changes in either supply or demand). We've quantified these problems, assuming that Britain had roughly 33 GW of wind power. To cope with lulls, we must effectively store up roughly 1200 GWh [gigawatt-hours] of energy (20 kWh per person). The slew rate we must cope with is 6.5 GW per hour (or 0.1 kW [kilowatts] per hour per person). There are two solutions, both of which could scale up to solve these problems. The first solution is a centralized solution, and the second is decentralized. The first solution stores up energy, then copes with fluctuations by turning on and off a *source* powered from the energy store. The second solution works by turning on and off a piece of *demand*.[12]

Other studies using this perspective examine technological innovation as a consequence of the geographic concentration of companies, or of differences in national "technological styles." Research on regional development shows that industries develop faster and are more successful when companies are clustered together, because spatial proximity stimulates information exchange and cooperation, which results in comparative advantages that are external to individual firms (Porter 1990; Powell 1990; Nitin and Eccles 1992; Locke 1994; Chandler 2001; Sorenson and Audia 2000; Murtha, Lenway, and Hart 2001). Clusters of firms result in booming regional industries, particularly when they form a decentralized regional network-based system (Saxenian 2000). In the case of the wind energy industry, studies have shown that geographic proximity and differences in technological approach account for the success of Danish wind turbine manufacturers compared to American manufacturers (Karnøe, Kristensen, and Andersen 1999; Garud and Karnøe 2003). Other studies have compared technological approaches and industry development in different countries such as the United States, Germany, and Denmark (Heymann 1998); the United States and Denmark (Est 1999); Germany, the Netherlands, and Sweden (Johnson and Jacobsson 2000); and the Netherlands and Denmark (Boon 2008).

Rather than focus on technology, numerous studies of new industries employ an economic or market perspective. According to this approach, the growth of a new industry such as the wind energy industry is determined first and foremost by market forces: the adoption of renewable energy policies, the deregulation of electric utilities, or the existence of bottlenecks in the supply of wind turbine components. For example, research on industry creation examines the way in which national policies influence industry emergence and development (Van de Ven and Garud 1989; Roy 1997; Freeland 2001; Murmann 2003; Murtha, Lenway, and Hart 2001; Ruttan 2001). In this view, different policy regimes produce different business strategies. Public policies create constraints and incentives for corporate actors and determine the level of competition in various organizational fields and industries (Edelman 1990; Baum and Oliver 1992; Davis, Diekmann, and Tinsley 1994; Fligstein 1990, 1996; Sutton and Dobin 1996; Dobbin and Dowd 1997, 2000).

A strength of the market perspective is that it shows how governments can stimulate the growth of the wind energy industry by adopting certain public policies such as power purchase agreements, investment and production incentives, and renewable set-asides (Redlinger, Andersen, and Morthorst 2002; Lauber 2005). Scholars who use this perspective analyze how energy markets are shaped not only by regulatory pressures but also by deregulation. Public policies sometimes shift the electricity market from a monopoly to a purchasing agency, in which competing electricity generators sell to a common purchasing agent; or to a competition model, in which generators compete to sell power to wholesale or retail customers either through bilateral contracts or power pools. In wholesale and retail competition markets, the price of transmission is crucial to the ability of renewable energy producers to compete. For this reason, many economic studies of deregulated electricity markets attempt to identify the optimal incentives for reducing transmission costs and increasing transmission capacity (Hunt and Shuttleworth 1996; Zaccour 1998; Griffin and Puller 2005; Lévêque 2006; Sioshansi 2008).

These technological and market perspectives dominate not only academic studies but also the public debate on the growth of the wind energy industry. Some commentators have criticized the wind energy industry for its technological or market shortcomings. Detractors have argued that wind power is not a major player in the energy sector and cannot become one in the foreseeable future because wind turbine technology is immature and the industry cannot compete without government subsidies. A 2006 *New York Times* article makes this argument forcefully:

Wind...generates a big problem: because it is unpredictable and often fails to blow when electricity is most needed, wind is not reliable enough to assure supplies for an electric grid that must be prepared to deliver power to everybody who wants it—even when it is in greatest demand....If wind machines reach 20 percent of total generating capacity, the cost of standby generators will reach $8 a megawatt-hour of wind. That is on top of a generating cost of $50 or $60 a megawatt-hour, after including a federal tax credit of $18 a megawatt-hour....Without major advances in ways to store large quantities of electricity or big changes in the way regional power grids are organized, wind may run up against its practical limits sooner than expected.[13]

Wind energy supporters have also used a technological and market framework to make predictions about the industry. Consider the arguments in a report produced by the Energy Watch Group in 2008:

1. The primary energy (wind) is cost-free;
2. The primary energy is renewable and never runs out;
3. There is an abundant resource, nobody can cut access/supply;
4. Stable life-cycle-cost of its use can be guaranteed;
5. Wind power is competitive with other new power sources;
6. Operating wind turbines cause no carbon emissions, no air pollution and no hazardous waste;
7. No water for cooling is needed;
8. Wind has a short energy payback of energy invested, normally less than one year;
9. There is a global, easy access to wind technology, compared to nuclear and others;
10. Time to market is very short, erection of entire wind farms within one year possible;
11. Fast innovation cycles prevail, based on maturing know-how;
12. Wind is still a young technology, allowing progress on the learning curve and cost reductions;
13. Wind is decentralized power; it allows small organizations or groups in various places to become a part of the power generation business and to sell it for a profit—very different from the exclusive structure of the oil, gas or nuclear business;
14. Distances from good wind sites to consumers in general are moderate (1–1000 miles) compared to other energy sources (oil, gas, uranium, coal);
15. Wind energy has positive side benefits for various stakeholders such as job creation, taxes, income options for farmers, infrastructure for remote areas, investment opportunities for local communities etc.;
16. Wind energy replaces expenses for (often imported) fuels by technology, creating energy, know-how and human labor in a decentralized way.

For these reasons we express the central thesis of this essay: *High growth rates of wind power generation worldwide will persist and wind power will conquer a large part of the energy market in the close foreseeable future.*[14]

Research focusing on wind energy technology and economics identifies a number of factors that shape the growth of the wind energy industry. However, it also leaves unanswered or does not consider a number of important questions: What drives the search for technological innovation? What contributes to the adoption of various renewable energy policies? Does the adoption of pro-wind policies guarantee their successful implementation? More generally, if wind power is not economically viable and the technology of wind turbines is inferior to that of conventional power plants, as detractors argue, why is the wind energy industry well developed in some countries and regions? And if wind energy is superior to fossil-fuel and nuclear energy, as supporters argue, why is the industry underdeveloped in many countries and regions?

This book argues that the success of the wind energy industry does not depend solely on technological or market design innovations; it also depends on the way in which societies respond to these innovations. As one author notes, "History judges a technology not on any absolute scientific criteria but on whether people used it and benefited from it" (Douthwaite 2002, 75). Many technological or policy innovations fail to diffuse widely and to contribute to new industries not because they are undeveloped, but because they are perceived as unnecessary or undesirable. However, relatively immature technologies and inefficient policies can diffuse rapidly and result in new industries if societies embrace them. Indeed, the history of the wind energy industry illustrates well both points. On the one hand, wind power technology was welcomed by Californians in the early 1980s even though the technology was immature and the life span of first-generation wind turbines averaged less than one year (Asmus 2001, 75). Similarly, an investment tax credits policy was used widely during the early 1980s despite the fact that oftentimes it did not result in electricity production but in tax shelters for investors (Redlinger, Andersen, and Morthorst 2002). On the other hand, wind power technology is opposed today in some regions with great wind potential even though the technology is proven—countries such as the United States and the United Kingdom have recently experienced significant local resistance to wind farms based on arguments about aesthetics.

Social Movements and the Wind Energy Industry

An alternative to the two perspectives briefly described above focuses on social movements or contentious politics. This perspective is uncommon; in fact, as one author notes, given the pervasive influence of social movements on the evolution of modern energy systems, it is surprising that the literature on the energy sector "has so often treated activists as irrelevant or passive agents" (Podobnik 2006, 13). While acknowledging the importance of technological or economic factors for the emergence of new industries, this approach holds that social movements can also play an essential role in industry development. Drawing on this perspective, I argue that to understand the global development of the wind energy industry, it is crucial to examine how environmental movements affect governmental policies, change the practices of electric utilities, and influence the behavior of organizational and individual electricity consumers.

Social movement scholars have long argued that movements contribute to social change in a number of ways. Social movements contribute to the adoption of public policies, particularly when the movements have well-developed organizational infrastructures (Soule et al. 1999; Cress and Snow 2000; Andrews 2001). Movements can shape the policymaking process not only through the intensity of their mobilization efforts but also via the particular political context or political opportunity structure in which they operate (Kriesi et al. 1995; Amenta and Young 1999; Burstein 1999; Soule et al. 1999; Andrews 2001; McCammon et al. 2001; Amenta and Caren 2004). This influence on the policymaking process is not as much direct—through a movement's mobilization efforts—as it is mediated by political context and public opinion (Cress and Snow 2000; Soule and Olzak 2004; Giugni 2007).

Movements can also lead to cultural and organizational change. Movements contribute to the construction of new social norms and values that shape broader cultural-change processes. Social movement organizers connect social movement issues with dominant cultural themes and produce "interpretative packages," or arguments that identify problems and explain how they can be rectified (Melluci 1989; Eyerman and Jamison 1991; Earl 2004). Because movements often develop alternative worldviews and attempt to spread them to the general public through direct actions as well as through mass media, the media serves a critical role in the diffusion of movement claims (Diani 2004; Oliver and Myers 2003; Hunt and Benford 2004). A growing number of studies show that social movement activists can push for change by working within organizations (Lounsbury 2001; Schneiberg 2002; Scully and Segal 2002; Lounsbury, Ventresca, and Hirsch 2003; Raeburn 2004), by pressuring them from outside (King and Soule 2007; King 2008), or by working from both inside and outside (Zald, Morrill, and Rao 2005; Rao 2009; Soule 2009).

While most social movement research centers on public policies or on cultural and organizational change, a few studies also point out that movements can affect entire industries by changing the norms that govern economic activities in sectors such as recycling (Lounsbury, Ventresca, and Hirsch 2003), electronics (Smith, Sonnenfeld, and Pellow 2006), and energy (Hess 2007). More recently, studies of the emergence of the wind energy industry in the United States as well as worldwide find that the environmental movement has a significant impact on industry growth (Sine and Lee 2009; Vasi 2009).

A social movement perspective on the global development of the wind energy industry has to start from the observation that the energy sector has been a site of contention for many years. The coal industry has experienced tremors because of labor activism in Europe and North America since the late nineteenth century, and the oil industry was targeted by nationalist movements that struggled to retain control over domestic oil industries particularly during the 1970s (Podobnik 2006). In our times, however, the most important challenges to the traditional model of energy production and consumption come from the environmental movement (Hirsh 1999; Podobnik 2006).

The Environmental Movement and the Electricity Sector

An often-quoted environmentalist adage states that "the Stone Age did not end because we ran out of stone, and the Fossil Fuel Age will not end because we run out of fossil fuels." Fossil fuels presently account for about 90 percent of world energy consumption, and they could continue to do so for the foreseeable future: it is estimated that at current production levels, coal will be available for at least the next 155 years, gas for at least 65 years, and oil for at least 40 years.[15] It would be naïve to assume that the recent growth of wind power and other renewable energy industries is due only to a growing concern that the world will run out of fossil fuels; however, it would be reasonable to assume that much of this growth is due to mounting concerns about the environmental impacts of fossil fuels. Environmentalists hope that, just as the Stone Age did not end because we ran out of stone but because bronze and iron were perceived as better substitutes, so too the Fossil Fuel Age will be replaced by a Renewable Energy Age before we run out of fossil fuels. To find better alternatives to fossil fuels, renewable energy industries need to grow and develop, and it's important that we understand what might drive such expansions. To specifically understand why the wind energy industry has developed unevenly worldwide, it is essential to examine not only the geographical variations in the distribution of natural resources for energy, but also the social variations in environmentalism.

The use of wind as an energy source has been intimately linked to geography for thousands of years. Given that wind is not as reliable as other sources of energy, wind power has been an attractive substitute in areas that have both good wind potential and poor conventional energy resources. Watermills were the dominant source of power for local economies in medieval Europe, but windmills dotted the medieval landscape in areas that suffered from drought or from a shortage of surface water—as in regions of

Spain—and in low-lying areas where rivers offered little energy, such as the lowlands of England and the Netherlands.[16] In our days, some countries with few fossil-fuel or hydropower resources—Spain again is an example—have developed a robust wind energy industry.

The use of various sources of energy to power the economy has been a contentious issue for many years. As far back as the twelfth century, religious authorities and aristocrats in England opposed the use of the post windmill by agrarian small businesses because the elites enjoyed a monopoly on the grinding of grain by their control of access to waterways and the use of the waterwheel. Realizing that wind is free and its use could not be controlled, the new entrepreneurs of the middle class built numerous windmills throughout the English countryside, causing many litigious disputes (Kealey 1987). Today, however, the contention over energy is generated mainly by the issue of the impact of energy production on the natural environment (Hirsh 1999; Podobnik 2006).

The harvesting and processing of fossil fuels—in particular coal mining methods such as mountaintop removal and strip mining, and offshore oil drilling and transportation—can create serious environmental problems. Combustion of fossil fuels results not only in sulfuric, carbonic, and nitric acids (which fall to earth as acid rain) but also in the release of significant amounts of radioactive materials such as uranium and thorium. In fact, one study has found that "Americans living near coal power plants are exposed to higher radiation doses than those living near nuclear power plants that meet government regulations."[17] Moreover, the burning of fossil fuels accounts for most of the anthropogenic emissions of greenhouse gases and contributes to global climate change.[18] Nuclear energy has its own environmental problems associated with uranium mining, processing, and transportation, as well as with the storage of spent nuclear fuel.

Environmental problems caused by fossil-fuel and nuclear energy production exist in virtually all industrialized and semi-industrialized countries; yet the public is often either unaware of these problems or unwilling to address them. As research by environmental sociologists has shown, environmental problems are socially constructed. In other words, they are defined and addressed by social, cultural, and political processes (Hannigan 1995). While the natural resources available in any region matter for the emergence of new industries, they frequently matter in a way that depends to a large degree on the practices and perspectives that are taken for granted in a given time and place (Freudenburg, Frickel, and Gramling 1995; Fisher 2006). As an example, when it comes to the adoption of environmental protection practices, the social perception of environmental degradation is often more important than the environmental degradation itself (Meyer et al. 1997; Frank, Hironaka, and Schofer 2000; Vasi 2006). Therefore, the wind energy industry is likely to develop when people no longer take for granted fossil fuels or nuclear power; in other words, when the public becomes both aware of the environmental problems associated with "dirty" energy and committed to address those problems by supporting the use of renewable energy. It is this book's contention that the public's perception of the pollution generated by the electricity sector is significantly shaped by the environmental movement.

I take the environmental movement to be defined as "a loose network of informal interactions that may include, as well as individuals and groups who have no organizational affiliation, organizations of varying degrees of formality that are engaged in collective action motivated by shared identity or concern about environmental issues" (Rootes 2004, 610). Although a number of scholars have argued that environmentalism has become "an interest group community" (Bosso 2000, 2005; Jamison 2001), there is no reason that environmental movement organizations and environmental interest groups cannot coexist. As Rootes (2004, 611) notes, "The balance of environmental movement actions has shifted from highly visible protests to lobbying and 'constructive engagement' with governments and corporations, much of which is publicly invisible but which, no less than more public forms of protest, contests established economic and social relationships and cultural understandings."

In this book I show that the growth of the wind energy industry is influenced by the environmental movement through four pathways. First, environmental activists, organizations, and research institutes contribute to energy policymakers' decisions to adopt and implement pro–renewable energy policies. This happens because the environmentalists develop alternative energy scenarios, lobby state and regional governments to adopt pro–renewable energy policies, and defend the implementation of these policies when they are under attack from the anti–renewable energy lobby. Second, the environmental movement changes energy consumers' preferences by raising awareness of the problems associated with fossil fuels and nuclear energy among residential and nonresidential customers of electric utilities. Third, the environmental movement contributes to the transformation of the electricity sector by changing energy professionals' values and attitudes regarding both conventional and renewable energy. Fourth, environmental groups contribute to the adoption of international environmental agreements such as the Kyoto Protocol, which result in the transfer of renewable energy technology. This book also recognizes that the growth of the wind power industry is influenced by other social forces, such as energy shortages (the energy crisis of 1973 is an example) and the nationalism of oil-exporting countries, as well as by economic forces, such as the need for electrification in developing countries. Before describing the book's structure, however, it is important to examine the evolution of wind power technology.

The Evolution of Wind Power Technology

Understanding how wind energy works in theory is relatively easy. Wind forms because the earth is unevenly heated by the sun—the poles receive less energy than the equator—and dry land heats up and cools down more quickly than oceans and lakes do. As the air moves between warm and cold regions, it forms a global atmospheric convection system reaching from the earth's surface to the stratosphere. In fact, scientists estimate that approximately 2 percent of the sunlight energy received by the earth is converted to the kinetic energy of the winds (Righter 1996, 3).

Wind power was initially used for transportation. As early as 3100 B.C. the Egyptians used sails for navigation up the Nile, while the Greek and Roman expansions around the Mediterranean Sea and beyond were based on wind power. During medieval times, wind was also used to power mills, particularly in areas where the use of watermills was impractical. Historians estimate that wind power provided as much as a quarter of Europe's energy needs between the fourteenth and nineteenth centuries, while hydropower and animal and human power provided the balance (Righter 1996, 15). In the United States alone, as many as six million water-pumping windmills were operating on the Great Plains and in the West between 1880 and 1930, making possible the expansion of the cattle industry (Righter 1996, 25).

While people have known for millennia that the kinetic power of the wind can be extracted by allowing it to blow past moving wings that exert torque on a rotor, it was only in 1920 that physicists determined the exact amount of power transferred to the rotor and the maximum amount of energy that can be extracted by a wind turbine. The amount of power transferred is directly proportional to the density of the air, the total area swept out by the blades, and the cube of the wind speed, while the maximum amount of energy that can be extracted by a wind turbine is 59 percent of the wind energy that flows past the turbine's blades.[19]

The first use of a windmill to generate electricity was a system built in 1888 by the American Charles Brush, and the first modern wind turbine was designed by the Dane Poul La Cour in 1891. Early twentieth-century wind turbines used modified propellers to drive direct-current generators and had a small electrical output of 1 to 3 kilowatts. These turbines were in widespread use in the rural areas of Denmark, Germany, and the United States. Experimental wind plants in the former USSR, the United States, Denmark, France, Germany, and Great Britain during the period 1930–1970 showed that large-scale wind turbines could work, but they did not result in the widescale manufacturing of practical, large wind turbines (Righter 1996).

Building small wind turbines is a relatively simple process, but producing significant amounts of electricity from wind is much more complex. Before 1985, many American designers and engineers focused on aerodynamic efficiency and low weight, assuming that the wind flows steadily past turbines. In reality, wind turbines operate in turbulent, unsteady winds that place tremendous stress on the turbine tower structure, blades, and gearbox. The American Wind Energy Association estimates that the power in wind may be five times greater at the height of the blade tip of a large wind turbine than at ground level, and that this wind can change direction and speed in a matter of seconds. Additionally, to be economically viable wind turbines have to operate for very long periods of time—more than 7,500 hours per year—with little or no maintenance.[20]

Wind power innovators and manufacturers have experimented with various turbine designs. Some innovators experimented with the vertical axis, or Darrieus, turbine, which resembles an upright egg whisk. Others experimented with horizontal axis turbines with one blade (monopteros), two blades, or three blades, either downwind or upwind. By the mid-1980s, however, most commercially available turbines utilized the proven "Danish concept," which features three blades facing into the wind.[21] Before the early 1990s the standard wind turbine used the Danish concept and also operated at fixed speed, meaning that the turbine's rotor speed was fixed and determined by the frequency of the supply grid, the gear ratio, and the generator design. This type of turbine had the advantage of being simple and robust, but had the disadvantage that fluctuations in wind speed were transmitted as fluctuations in mechanical torque and in electrical power on the grid. Starting in the mid-1990s, manufacturers began building variable-speed wind turbines. These turbines capture more energy, reduce mechanical stress on the turbine, and reduce fluctuations in electrical power on the grid, but use more components and have higher equipment costs.[22] Gradually, variable-speed turbines have become more and more popular because their advantages outweigh their disadvantages. As figure 3 shows, wind turbines with variable speed and partial- or full-scale frequency converters (types C and D) accounted for almost 43 percent of the world market share in 1998, but their share increased to over 67 percent in 2002.[23]

The birth of the modern wind energy industry can be traced to the early 1980s. As new materials and technologies became available, the reliability of wind turbines dramatically increased. As figure 4 shows, by 1995 the reliability of wind turbines was better than that of a diesel generator, and by 2000 it approached the reliability of the combined cycle gas turbine. Indeed, four stages can be identified in the development of the modern wind energy industry. The first is an "early-experimentation" stage, from the beginning of the twentieth century to 1970, characterized by very high energy costs and very low turbine reliability. The second is an "experimentation stage," from 1970 to 1990, characterized by high energy costs (an average cost of more than $0.10 per kWh) and low turbine reliability (a turbine failure rate significantly higher than the diesel generator failure rate). The third is an "early-consolidation stage," from 1990 to 2000, characterized by somewhat high energy costs (an average cost of $0.05–0.10 per kWh) and relatively good turbine reliability (a turbine failure rate similar to or better than the diesel generator failure rate). The final stage in the development of the modern wind energy industry is a "consolidation stage," from 2001 to the present, characterized by competitive energy costs (an average cost of $0.05 or less per kWh) and very good turbine reliability (a turbine failure rate better than

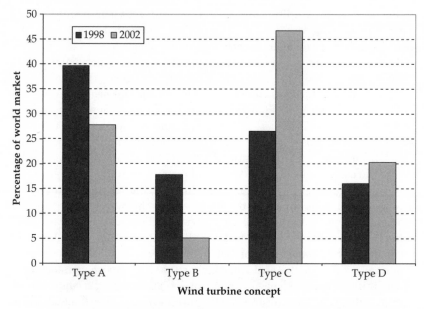

Figure I.3. World market share of wind turbine concepts in 1998 and 2002. Legend: Type A = fixed speed; Type B = limited variable speed with variable generator-rotor resistance; Type C = variable speed with partial-scale frequency converter; Type D = variable speed with full-scale frequency converter. Adapted from Hansen 2005.

the combined cycle gas turbine failure rate and approaching the steam turbine failure rate).

As the reliability of wind turbines constantly improved between 1981 and 2000, their size and power output also dramatically increased: the average size of the rotor diameter increased from 10 to 71 meters and the average rated capacity increased from 25 to 1650 kW.[24] Consequently, the cost of electricity from wind turbines dropped by more than 80 percent over approximately twenty years, from more than \$0.30 per kWh in the early 1980s to less than \$0.05 in 2005.[25] Figure 5 shows that the price of electricity obtained from wind at high wind-speed sites with good power grid access has become competitive with the price of electricity from conventional sources.

The main disadvantage of wind power is the intermittency of the wind. Because wind is not constant, most wind turbines have a capacity factor between 25 and 40 percent.[26] For comparison, fossil-fuel power plants have a capacity factor between 40 and 80 percent. This does not mean that most wind turbines run only 25 to 40 percent of the time. For example, wind turbines at typical locations in the Midwestern United States run from 65 to 90 percent of the time, but they generate at less than full capacity, making their capacity

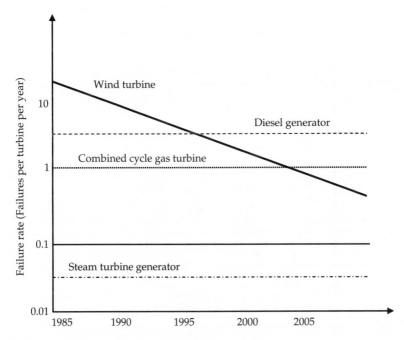

Figure I.4. Wind turbine reliability—failure rate over time in comparison with other power sources. *Windstats* 18, no. 1 (2005).

factor lower. Because available wind energy varies with the cube of the wind speed, a relatively small increase in wind speed will correspond to a large increase in available energy. Modern wind turbines begin to operate at wind speeds above 3 meters per second, and their power output rapidly increases between 5 and 11 meters per second. The turbines reach their rated capacity at wind speeds between 12 and 15 meters per second, depending on their individual design. Figure 6 shows the power curve of a modern 2.5 MW turbine manufactured by General Electric. This turbine reaches its rated capacity at wind speeds of approximately 12 meters per second. At very high wind speeds, above 22 or 25 meters per second, most wind turbines are designed to "spill" some of the energy available in the wind to avoid structural damage.[27] Therefore, wind turbines produce maximum power within a certain wind-speed interval that has an upper limit at the cut-out wind speed.

A number of strategies can be used to deal with the intermittency of wind. When the wind is very strong and the output of wind farms exceeds the local demand for electricity, excess energy can be stored by pumping water in a hydrostorage facility—a system of two artificial lakes at different altitudes. The stored energy can then be used when the electricity produced from wind is not enough to satisfy local demand. The problem of local storage can be

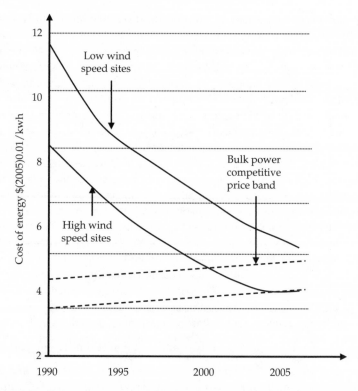

Figure I.5. Wind energy cost trends—leveled cost of energy in constant 2005 dollars. National Renewable Energy Laboratory Energy Analysis Office.

sidestepped if electricity can be easily transported between regions with abundant and cheap wind energy and regions with existing hydropower facilities that generate power from a natural inflow. Hydropower facilities create peak power at a very low cost because there are no pump losses. Hydropower is also very low cost during part-load operation and at start-up. Therefore, the low-cost solution for decreasing uncertainty in a power system due to wind fluctuation is to transport electricity between regions with wind farms and regions with hydropower plants—but this solution assumes that transmission lines are available. The volatility of wind power delivery can also be decreased by spreading wind farms over a large area, since the hourly correlation of wind variability drops dramatically with distance between sites.[28]

Despite the technological challenges posed by large-scale wind development, wind power has become an attractive option due to its many advantages over other sources of energy. Wind energy depends on a free and abundant fuel source, has one of the shortest energy-payback times of any energy technology, provides more jobs per dollar invested than any other energy technology, and has a price that is relatively immune to inflation.[29]

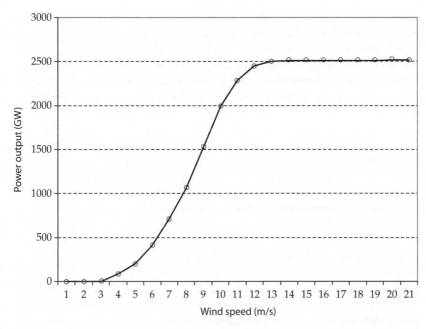

Figure I.6. The power curve of a 2.5 MW wind turbine—GE 2.5xl. Adapted from German Wind Energy Association, "Wind Energy Market 2007/2008."

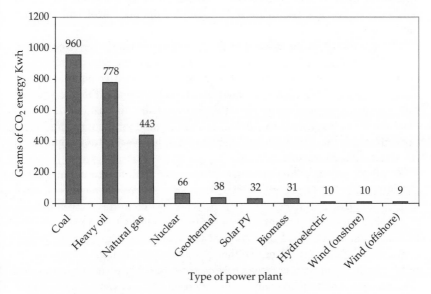

Figure I.7. Carbon dioxide emissions from various energy sources. Adapted from Sovacool 2008.

Additionally, figure I.7 shows that the production of electricity from wind turbines results in significantly fewer greenhouse gases than the production of electricity from coal, oil, natural gas, or nuclear power plants; in fact, wind power emits even less carbon dioxide per kilowatt-hour of electricity produced than solar photovoltaic (PV) and other renewable energy sources.[30]

Despite the popular belief that wind turbines are extremely dangerous for birds, studies show bird mortality caused by wind turbines is trivial when compared to bird mortality caused by other human activities. While wind turbines are responsible for less than 0.01 percent of annual avian mortalities, buildings account for over 58 percent, power lines for over 13 percent, cats for over 10 percent, and automobiles for over 8 percent.[31] It is not surprising that by the beginning of the twenty-first century, wind energy has developed into a professional, multi-billion-dollar, high-technology industry.[32]

Methodology and Plan of the Book

This book's main theoretical contribution is to bring social movements into the study of market formation and industry growth. The book identifies the main, "ideal type" pathways through which social movements contribute to industry development: top-down, by shaping the policymaking process; bottom-up, by creating consumer demand for a new product; midpoint, by changing the dominant rationale in an industrial sector; and unintended, by redistributing capital and technology at the transnational level. This book also identifies the various factors that mediate social movements' impact on emerging industries: political context, public opinion, mass media, and natural resources.

The empirical study presented here analyzes how the environmental movement matters for the development of the wind energy industry. This study uses a multimethod approach, combining quantitative and qualitative analyses. Multivariate regression analysis is used to identify the main factors that shape the growth of the wind energy industry in national contexts. Qualitative analyses are then used for in-depth examination of the processes through which the environmental movement influences the development of the wind energy industry.

More than seventy interviews were conducted with environmental activists and professionals, energy policymakers, and wind power engineers, innovators, and advocates over a five-year period in six countries.[33] Additionally, the qualitative research relies on existing documents such as newspaper and journal articles published over the last three decades. This approach overcomes some of the limitations of either quantitative or qualitative research. To put it simply, quantitative research is used for the "big picture": to develop generalizations and theorize about the environmental movement's role in the development of the wind energy industry. Qualitative research is then used for a "detailed film," that is, to enhance understanding of the dynamic and interactive processes through which environmental activists and organizations contribute to industry growth.

The first chapter of the book develops a theoretical model based on existing research on social movement outcomes and industry creation. It begins by identifying the environmental movement's main pathways of influence on the wind energy industry; next, it employs cross-national multivariate regression analysis to test the heuristic model.

The second and third chapters use numerous semistructured interviews with environmental activists, energy policymakers, and wind energy professionals to describe how the adoption and implementation of pro–wind energy policies is shaped by the environmental movement in different countries and regions. The second chapter focuses on the contentious processes surrounding the adoption and implementation of renewable energy feed-in tariffs (FITs) in countries that developed the first and most effective tariffs: Germany, Denmark, and Spain. The third chapter focuses on the contentious processes surrounding the adoption and implementation of a renewable portfolio standard (RPS) in countries that have very good wind potential but somewhat underdeveloped wind energy industries: the United States, Canada, and the United Kingdom.

The fourth chapter focuses on the role of environmental groups and activists in creating consumer demand for renewable energy. The chapter uses semistructured interviews to explore the hypothesis that the recent increase in consumer demand for renewable energy in countries such as the United States is caused mainly by environmental campaigns for renewable energy, in particular wind power.

The fifth chapter examines how environmental groups and activists can contribute to a shift in the electricity sector's rationale by changing energy professionals' values and attitudes regarding both conventional and renewable energy. The chapter uses interviews with policymakers, environmental activists, and energy professionals from different countries to explore the idea that the electricity sector's rationale changes gradually when environmental activists and sympathizers are able to gain control of energy professional societies, critique the traditional logic of energy production, and offer a solution—hinging on an environmentalist logic—to the electricity sector's problems.

The conclusion argues that environmentalist "global winds of change" are almost as important as the atmospheric winds for the development of the wind energy industry around the world. It briefly examines the unintended consequences of the environmental movement, and focuses on the transfer of wind power technology from developed to developing countries. It also describes the role of other factors that contribute to the global development of this industry, and presents a few implications for future studies of industry creation as well as energy sector growth.

1

The Big Picture

The Environmental Movement's Impact on the Global Development of the Wind Energy Industry

Where there is wind, there is turbulence, friction, and ultimately, conflicts. On the physical plane, winds are all about chaos and particles of matter so minute they escape detection, yet they possess incredible amounts of energy. Chaos also dominates humanity's efforts to harvest the power of wind.
—Peter Asmus, *Reaping the Wind: How Mechanical Wizards, Visionaries, and Profiteers Helped Shape Our Energy Future* (Washington, D.C.: Island Press, 2001)

A Framework for Analyzing the Global Development of the Wind Energy Industry

In 1996 wind power became the world's fastest-growing energy source, reaching an annual growth rate of 32 percent. An article published at that time by the environmental think tank Worldwatch Institute noted that wind power had had an annual growth of 20 percent since 1990, while nuclear power and coal combustion had grown at a rate of less than 1 percent per year.[1] Comparing the renewable energy industries and the high-tech industries, the article argued that "the computer industry has shown the powerful effect of double digit growth rates. The fact that personal computers provided less than 1 percent of world computing power in 1980 did not prevent them from dominating the industry a decade later. Already, wind power went from providing less than 1 percent of the electricity in the north German state of Schleswig-Holstein in 1990 to 8 percent in 1995."[2] The article emphasized that wind power had the potential to exceed 20 percent of world electricity in the near future, and it concluded that, in the long term, "wind power could...eventually replace coal and nuclear power and allow a sharp reduction in world carbon emissions."

Wind power advocates have argued for a long time that the wind energy industry could—and should—produce a significant proportion of electricity worldwide. Their argument is that wind will inevitably become a major source of electricity production because constant technological innovation will lead to significant declines in the price of wind power. Over the last fifteen years, the cost of manufacturing wind turbines has declined by approximately 20 percent each time the number of manufactured turbines has doubled, and the production of large-scale turbines has doubled every three years. The Danish Energy Agency predicted in 1996 that a further cost reduction of 50 percent could be achieved by 2020 (Ackermann 2005, 18).

Worldwide, however, wind power remains somewhat underdeveloped. By the end of 2008, only sixty-eight countries had a multi-megawatt wind energy industry and only thirty-two countries had more than 100 MW of wind energy capacity. As figure 1.1 shows, wind energy production was concentrated in Europe (Denmark, Spain, and Germany), North America (the United States), and China and India. In terms of the percentage of electricity produced from wind, only three countries (Denmark, Spain, and Germany) produced more than 5 percent of their electricity from wind at the end of 2008 (see figure 1.2).

Scholars who examine the past development of the wind energy industry and estimate its future growth focus primarily on technological or economic factors. Many energy professionals have focused on maximizing wind resources by improving wind forecasts and eliminating bottlenecks in electricity transmission (Gipe 1995; Redlinger, Andersen, and Morthorst 2002; Ackermann 2005; Bhadra, Kastha, and Banerjee 2005; Heier 2006; MacKay 2008). Many economists have also analyzed the role of market forces such as energy-related policies, the deregulation of electric utilities, the supply and demand of wind turbine components, and various transmission issues (Hunt and Shuttleworth 1996; Zaccour 1998; Griffin and Puller 2005; Lévêque 2006; Sioshansi 2008). Yet, given the pervasive influence of social movements on the evolution of modern energy systems, it is surprising that only a few researchers have examined the contention over energy production and consumption, or the way in which the environmental movement shapes this industry (Podobnik 2006; Sine and Lee 2009; Vasi 2009).

This chapter presents a contentious politics or social movements account of the global development of the wind energy industry. It recognizes that technological and market forces have contributed to the fact that today wind power is a multi-billion-dollar industry with tremendous prospects for future growth. But these two dominant interpretations overemphasize the role of technological and economic factors, downplay the role of social and political factors, and are unable to fully account for the variable global development of the wind energy industry. The chapter has two main goals: to develop a model that presents the big picture of the environmental

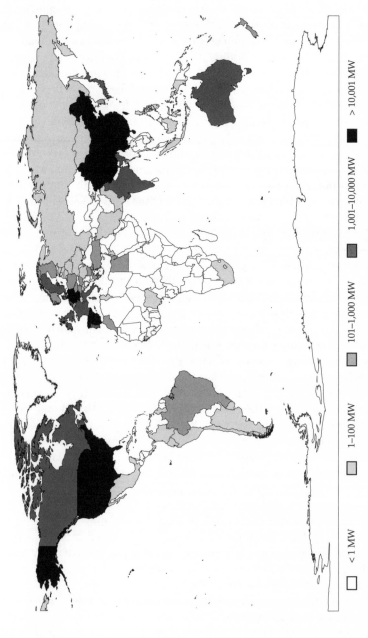

Figure 1.1. Global distribution of wind energy production by country at the end of 2008 (five categories; darker shades indicate higher total-installed wind power capacity).

Legend:
- ☐ <1 MW
- 1–100 MW
- 101–1,000 MW
- 1,001–10,000 MW
- > 10,001 MW

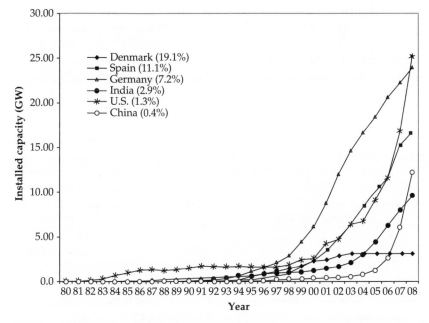

Figure 1.2. Cumulative installed wind power capacity by selected country 1980–2008; the percentage of electricity produced from wind at the end of 2008 is indicated in parentheses. Adapted from Earth Policy Institute and World Wind Energy Association.

movement's pathways of influence on the wind energy industry, and to test this model using quantitative analysis.

Research on Industry Creation and the Wind Energy Industry

Why are new industries such as the wind energy industry more developed in some countries and regions than others? Answers to this question vary with theoretical perspective and methodological approach. A number of studies of industry emergence and development employ a technological perspective. Some of these studies focus on the effect of contextual factors such as the availability of material and organizational resources for technological innovation and industry emergence. The availability of material, or natural, resources contributes to the emergence of new industries and to their subsequent geographic concentration because industries located in areas with rich natural resources minimize transportation costs. The steel industry formed and concentrated near iron deposits, while the oil refining industry concentrated near areas with crude oil deposits and access to transportation routes (Harris 1954; Chernow 1998). In the case of wind power, wind turbine manufacturing facilities and wind farms were initially located in areas with good wind potential and in relative proximity to power grids because these factors lower the cost of wind energy production and transmission (Gipe 1995; Bhadra, Kastha, and Banerjee 2005; Heier 2006). To take full advantage of high-quality wind and as noted above, energy professionals seek to improve wind forecasts and eliminate bottlenecks in electricity transmission (Redlinger, Andersen, and Morthorst 2002; Ackermann 2005; Heier 2006; MacKay 2008).

Organizational resources are also important for technological innovation and industry formation. Many industries are geographically clustered because they depend heavily on both formal and informal cooperation and information exchange (Powell 1990; Nitin and Eccles 1992; Locke 1994). Clusters of firms enjoy a competitive advantage due to external economies— self-reinforcing agglomerations of technical skill, infrastructure, capital, suppliers and services that result from the informal flow of information and proximity to universities (Porter 1990). Some studies compare industry growth to pollination, arguing that "the presence of many structurally equivalent organizations increases the pool of potential entrepreneurs in a manner similar to a pollination process in which plants produce pollen that blows away in the wind only to land somewhere nearby and burst into new plants" (Sorenson and Audia 2000, 442). For example, the development of high-tech industries manufacturing semiconductors or flat-panel displays has been heavily influenced by the geographic concentration of organizations that have interacted frequently (Chandler 2001; Murtha, Lenway, and Hart 2001). In this model, industries grow because knowledge accumulates and "spills over" when other manufacturers are close by, not because governments or other agencies manage industry creation.[3]

Other studies of regional industrial development, however, argue that not all clusters of firms are the same. New industries are likely to be successful in locations where companies have a "regional advantage" due to a specific industrial system characterized by shared identities, high levels of trust, and increased competitive rivalries. For example, while a firm-based industrial system such as Massachusetts' Route 128 is constrained by the isolation of its producers from external sources of know-how and information, a network-based industrial system such as Silicon Valley can adapt better to new technologies because it supports experimentation and learning (Saxenian 2000). Consequently, while the Silicon Valley cluster enhanced its vitality during the 1980s, the "Massachusetts Miracle" of Route 128 ended abruptly in the same period.

In the case of the wind energy industry, research shows that the early success of the Danish wind turbine manufacturers is attributable not only to the availability of strong winds in Denmark but also to the geographical concentration of start-up companies that adopted a specific technological approach. Danish manufactures used a "bricolage approach," a collective and mutually adaptive process in which networked firms proceeded through small gains to gradually improve wind turbine technology (Karnøe, Kristensen, and Andersen 1999; Garud and Karnøe 2003). This particular approach contributed to the Danish domination of the global market for wind turbines even though the U.S. and German governments spent significantly more than the Danish government on wind power research and development between 1975 and 1988 (Heymann 1998).[4] In contrast, American manufacturers used a "breakthrough approach," a process in which individual firms compete to achieve a technological innovation in one great leap. Because these firms exchanged little information, wind turbines that performed well under controlled laboratory conditions had high failure rates in turbulent, real-world atmospheric conditions. Designed to operate over a twenty-year period, first-generation wind turbines installed in California averaged less than ten hours in actual life span (Asmus 2001, 75). Thus, studies adopting a technological perspective conclude that wind energy industries are strong when countries have high-quality wind and use a specific technological approach to wind turbine development and manufacturing.

Technological reasons alone, however, cannot explain industry development—there must be a market for the product, and the industry must be an economically viable activity. Studies taking a market perspective argue that governments contribute to the emergence of new markets through public policies. While governments cannot command businesses to perform, they can adopt new policies that establish the rules of competition and cause firms to adopt or invent new practices (Meyer and Rowan 1977; DiMaggio and Powell 1983; Edelman 1990; Sutton and Dobbin 1996). The most important role of public policies is to create constraints and incentives for corporate actors and to determine the level of competition in various industries (Fligstein 1996; Dobbin and Dowd 1997, 2000; Roy 1997; Russo 2001). Indeed, research shows that the success of new firms in various emerging industries is often determined by the type of public policies that frame the competitive environment (Fligstein 1990; Davis, Diekmann, and Tinsley 1994; Baum and Oliver 1992).

The wind energy industry is no exception; it is heavily influenced by the adoption of particular energy policies. Power purchase agreements contribute to growth by mandating that utilities purchase independently generated power at predetermined costs. Other beneficial policies include investment incentives, which reduce wind-power-project capital costs by offering developers governmental subsidies, and production incentives, which offer subsidies per kilowatt of electricity generated. Still other important policies include environmental taxation, which adds to the cost of fossil-fuel-generated energy by imposing taxes on pollutants, and renewable set-asides, which create separate markets for competition among producers of green energy by mandating that a certain percentage of electricity come from renewable sources (Redlinger, Andersen, and Morthorst 2002).

Deregulation and transmission price also shape energy markets and the growth of the wind power industry. Deregulation can shift the electricity market from a monopoly to either a purchasing agency (when competing electricity generators sell to a common purchasing agent), or to wholesale or retail competition (when electricity generators compete to sell power to wholesale or retail customers). In deregulated markets, the price of transmission is also crucial to the ability of wind power producers to compete, as transmission is usually unbundled from generation in wholesale and retail competition markets. Therefore, some economic studies of deregulated electricity markets search for incentives that could reduce the costs of wind energy transmission (Hunt and Shuttleworth 1996; Zaccour 1998; Griffin and Puller 2005; Lévêque 2006; Sioshansi 2008).

While these perspectives offer important insights into the growth of the wind energy industry, they have a number of shortcomings. Studies that focus on the role of material resources often fail to recognize that the perception of the availability of resources is socially constructed. High-quality wind is an important precondition for the emergence of the wind power industry in a specific locality, but local residents often perceive strong wind as an unavoidable nuisance of nature—rather than as a resource for producing electricity—if they have not been exposed to information provided by wind power supporters. For example, most farmers in Texas did not think of wind as a resource for producing electricity until renewable energy advocates organized educational campaigns and distributed information at county fairs and town meetings.[5]

Similarly, studies that emphasize organizational resources such as the geographic proximity of structurally equivalent organizations are better suited to explain why the wind energy industry is concentrated in certain regions or why certain wind turbine manufacturing centers perform better than others, rather than to explain why the industry develops globally or in geographically distant locations. If the process of industry emergence is similar to a pollination process in which companies act as plants and entrepreneurs act as pollen produced by the companies and "blown away in the wind" to land nearby and create new companies, what accounts for the force, direction, and speed of "the wind"?[6] More specifically, why is it that the wind energy industry has developed even in countries and regions that do not have a high density of manufacturers of wind turbine components?[7]

Studies that focus on the role of public policies rarely address the issue of their origin and implementation. Research on the growth of the wind energy industry conducted under this framework shows that the adoption of specific energy policies contributes to the development of the industry, but it does not examine the factors that contribute to the adoption of favorable policies. Put another way, if the adoption of pro–wind energy policies contributes to the development of the wind energy industry, what contributes to the adoption of these policies? Moreover, the adoption of renewable energy policies does not guarantee their successful implementation. The adoption of the Public Utility Regulatory Policies Act (PURPA) in 1978 did not result in sustained growth for the renewable energy industries in the United States because its implementation was left to individual states.[8] These shortcomings can be addressed by building on research on social movements and contentious politics.

Social Movement Outcomes, the Environmental Movement, and the Wind Energy Industry

To address the limitations inherent in the technological and economic perspectives on industry emergence, it is necessary to examine research on the intended as well as unintended consequences of social movements.[9] Much research on social movement outcomes focuses on movements' effect on public policy through institutionalized tactics such as litigation and lobbying, and the research shows that the success of these tactics hinges on organizational capacity. The stronger the infrastructure of the movement organization, the more likely it is that policies relevant to movement goals are going to be adopted and implemented (Soule et al. 1999; Cress and Snow 2000; Andrews 2001).

Other studies focus on the role of political opportunity structures, or on the "consistent, but not necessarily formal or permanent, dimensions of the political environment that provide incentives for people to undertake collective action by affecting their expectations for success or failure" (Tarrow 1994, 85). Political opportunities are external resources that can benefit even weak and disorganized challengers. For example, some movement mobilization and policy outcomes are influenced by the presence of elite allies or by changes in the level of elite receptivity to protests or elite willingness to repress protests (Kriesi et al. 1995; Amenta and Young 1999; Burstein 1999; Soule et al. 1999; Andrews 2001; McCammon et al. 2001; Amenta and Caren 2004). Factors such as the presence of political allies have a major influence on movement outcomes independent of the level of movement mobilization.

Still other studies argue that social movements' impact on public policy is interactive and contingent, as movements have not only direct and indirect effects—through mobilization, political alliances, and favorable public opinion—but also "joint-effects" (Cress and Snow 2000; Soule and Olzak 2004; Giugni 2007). Most policymakers do not engage in substantial policy reform without a strong and supportive social movement as well as strong political allies or favorable public opinion, or both (Giugni 2007). Moreover, a movement's ability to influence policy depends on the kind of issue or policy areas it addresses. In certain areas, authorities have a limited margin for action; in others, authorities perceive a serious threat from social movement demands and a direct challenge of their power.[10]

Broad cultural change is another outcome of social movements through the construction of new social norms and values (Melluci 1989; Eyerman and Jamison 1991; Earl 2004). Social movement organizations are capable of mobilizing their bases, attracting new members, and changing the public's social values because they use various frame alignment strategies—a process in which the interpretive orientations of individuals and social movement organizations are linked "such that some set of individual interests, values and beliefs and SMO [social movement organization] activities, goals, and ideology are congruent and complementary" (Snow et al. 1986, 464). Key— also called "critical"—communities of social movement actors constantly identify problems and explain how they can be rectified, at the same time that they connect social movement issues with dominant cultural themes (d'Anjou and Van Male 1998; Rochon 1998). While these key communities develop alternative worldviews, social movements form around these alternative views and advocate them to the wider public through mass media and political action (Snow et al. 1986; Klandermans, Kriesi, and Tarrow 1988; Gerhards and Rucht 1992). Mass media serves a vital role, as a mediator between movements and the public, in the diffusion of movement claims and in the mobilization of sympathizers for specific campaigns (Diani 2004; Oliver and Myers 2003; Hunt and Benford 2004).

Finally, studies of social movement outcomes have also focused on organizational change. Movements can affect organizations' decisions to adopt new practices and change their policies in different ways. Individuals who are social movement activists or sympathizers and who also work for an organization can act as internal agents of change by finding new solutions to collective problems and mobilizing resources for change within the organization (Lounsbury 2001; Schneiberg 2002; Scully and Segal 2002; Lounsbury, Ventresca, and Hirsch 2003; Raeburn 2004). Social movement activists can also pressure organizations to change from outside by organizing demonstrations, protests, boycotts, and lawsuits (Zald, Morrill, and Rao 2005; King and Soule 2007; King 2008; Soule 2009). As Rao (2009, 7) observes, activists who work inside and outside organizations are able to reshape markets if they can "forge a collective identity and mobilize support by articulating a *hot cause* that arouses emotion and creates a community of members, and relying on *cool mobilization* that signals the identity of community members and sustains their commitment."

One area where very little research exists on the influence of social movements is new and emerging industries. A few studies do point out that movements can influence the development of economic activities such as the recycling industry (Lounsbury, Ventresca, and Hirsch 2003), the electronics industry (Smith, Sonnenfeld, and Pellow 2006), and the energy sector (Hess 2007; Podobnik 2006). More recently, a study of the emergence of the wind energy industry in the United States argues that environmental organizations aided entrepreneurial activity by interpreting events as opportunities for collective action, distributing information about new forms of energy production, and mobilizing their memberships to support the wind power sector (Sine and Lee 2009). Similarly, another study finds that the environmental movement shapes the development of the wind energy industry at both national and subnational levels (Vasi 2009).

The energy sector arouses frequent contention, and it is not possible to understand the development of the wind energy industry without insight from studies of social movement outcomes. In particular, it is essential to examine the role played by the environmental movement. The environmental movement is one of the most influential movements in the last fifty years: it has indelibly shaped public opinion, spurred the creation of green political parties, and contributed to the "greening" of everyday behaviors and organizational practices (Hoffman 1997; Almanzar, Sullivan-Catlin, and Deane 1998; Rucht 1999; Lounsbury 2001; Hoffman and Ventresca 2002; Rootes 2004).

It is no coincidence that the first signs of public concern about fossil-fuel and nuclear energy pollution, and support for alternative energy, correspond to the rise of the modern environmental movement in the late 1960s and early 1970s. With no public outcry or publicity, the electricity sector had been damaging the environment long before the emergence of the environmental movement, and prior to the 1970s the general public had little or no interest in clean, alternative energy. Although a significant number of small wind-powered electric generators had been installed in different parts of the world in the first half of the twentieth century, they were not perceived as a potential solution to the environmental pollution caused by fossil fuels.[11] Indeed, as Righter (1996, 141) notes, when engineers presented plans to the U.S. Congress in 1951 to build large-scale wind turbines,

> no government official or other (wind energy) advocate at the hearing played the pollution card. No one noted that by producing enough energy through wind power rather than oil or coal, a significant byproduct would be clean air. By the 1950s, air quality was deteriorating in such metropolitan areas as Los Angeles, but officials placed the blame solely on the automobile. It was too early for Americans to grasp the complexity of air pollution, or comprehend the difference between a clean plant and one that polluted.

The environmental movement's impact on the public perception of the electricity sector became evident in 1970, after the first Earth Day. In the United States, opinion polls surveying perception of major domestic problems in 1970 indicated that the problem of "reducing pollution of air and water" had moved to second place from a ranking of ninth in only five years (Righter 1996, 153). In 1971, Stuart Udall, the former secretary of the interior under President John F. Kennedy, described wind turbines in a newspaper interview as "symbols of sanity in a world that is increasingly hooked on machines with an inordinate hunger for fuel and a prodigious capacity to pollute."[12] In 1973, twenty-two years after the first presentation before Congress of the plans to build large-scale wind turbines, the United States Senate held hearings on a bill aimed at providing funds for research on electricity generation "with minimum impact upon the environment." In contrast to the 1951 hearings, an explicit connection was made between energy production and environmental pollution.[13] The environmental movement's campaigns also contributed to the mounting concern over nuclear energy throughout the 1970s and to a major turnaround in American utilities' plans to build nuclear plants. More than 230 nuclear plants had been planned by the end of 1974, but only 15 were planned afterward, and no plants have been completed since 1978.[14]

Public awareness of the pollution generated by fossil fuels and nuclear power continued to increase during the 1980s in Europe and North America. The issue of nuclear energy was often the centerpiece of environmental protests. In Germany almost 50 percent of the environmental protests between 1988 and 1997 were directed against nuclear energy (Rucht and Roose 2003). Numerous environmental groups became advocates of renewable energy, and in particular wind power, promoting it as the most ecologically benign source of electric power.[15] As environmental campaigns raised awareness of issues such as acid rain, air pollution, and nuclear waste, many nonprofit environmental organizations rapidly increased in membership. For example, the number of British members or supporting donors of Greenpeace grew tenfold between 1981 and 1991, and those of Friends of the Earth grew sixfold (Rootes 2003). Similarly, in Italy membership in six major environmental groups increased by 250 percent between 1983 and 1988, while in Germany the four largest environmental nongovernmental organizations (NGOs) experienced significant growth during the 1980s (Diani 1995; Rucht and Roose 2003). In the United States, the Sierra Club increased in membership from 181,000 to 560,000 between 1979 and 1990, while the ten largest environmental groups grew from approximately 1 million members in 1979 to almost 7 million in 1990 (Mitchell, Mertig, and Dunlap 1992; Van der Heijden 1999).

By the 1990s, the environmental movement was able not only to increase awareness of the environmental problems associated with energy production, but also to create the perception that renewable energy, in particular wind power, is a viable solution to these problems. As a result of these environmentalist "winds of change," in a few decades wind turbines underwent a dramatic transformation in the public mind: from "small machines for farms" to "large farms of machines." In other words, they went from a technology that could satisfy only isolated farmers' electricity demands to a technology that could satisfy the electricity demands of millions of people and also address the most pressing environmental problems of the world.

Two factors account for the environmental movement's enhanced ability to change energy-related values, attitudes, and behaviors since 1990. The first is that the movement became institutionalized and gradually increased in membership in many countries. The number of environmental movement organizations increased dramatically between 1970 and the mid-1990s. The number of environmental groups in the United States more than doubled, from approximately 200 to 500, and the number of transnational environmental groups rose sixfold in the same period from less than 30 to approximately 180 (Johnson and McCarthy 2005). In 2003, only human-rights transnational social movement organizations (TSMOs) were more numerous than environmental TSMOs (Smith 2008, 123). In the same year, the American environmental advocacy community totaled approximately 8 million members, and the combined revenue of the top thirty nonprofit environmental organizations was well above $2 billion (Bosso 2005, 7).

Environmental groups underwent a process of institutionalization both internally, through bureaucratization, professionalization, and specialization; and externally, through cooperation with governments and businesses, and through the adoption of nonconfrontational tactics (Van der Heijden 1999; Rootes 2003; Bosso 2005). In fact, the transformation of many U.S. environmental organizations from relatively unprofessional groups supported by a few elite patrons into mass-based, professional advocacy organizations has led some authors to conclude that the movement has become "a mature and very typical American interest-group community, albeit one with an impressive array of policy niches and potential forms of activism" (Bosso 2005, 157). However, critics of the institutionalization trend have argued that environmental organizations depend too much on foundation grants and corporate donations or on passive "checkbook" or "credit card" supporters of noncontroversial and nonconfrontational tactics (Mitchell, Dunlap, and Mertig 1992; Dowie 1995; Brulle 2000; Bosso 2005). Consequently, the environmental movement may contribute to the adoption of "green ceremonial facades," or well-meaning fictions that create the impression that an organization is green but, due to a lack of grounding in the deeper layers of an organization's culture, have little effect on the organization's actual environmental performance (Forbes and Jermier 2002, 206).

Regardless of the terminology used to describe the environmental movement's transformation, it is undeniable that by the early 1990s, environmental activism had created a relatively widespread public perception that humans should do more to protect the environment and "save the Earth." At the same time, major organizations such as Greenpeace and Friends of the Earth had become committed not only to critical campaigning but also to funding research on environmental issues and to "solutions campaigning" designed to promote better environmental practices (Rootes 2003, 22).

Another factor that accounts for the movement's enhanced ability to change energy-related values, attitudes, and behaviors is the rapid ascension of global climate change to the top of the environmental agenda. As with other global scientific formulations, climate change has been very attractive to the environmental movement because it permits the "packaging of multiple environmental problems and concerns within a common, overarching rubric," and it "conveys the legitimacy and persuasiveness afforded by being rooted in science" (Buttel and Taylor 1994, 242). Global climate change has become the focus of sustained campaigning as well as a consolidating framework for the environmental movement.

The environmental campaigns on global climate change have not only raised awareness of this problem but also created support for renewable energy. Surveys show that 70 percent of the American public in 1995 considered the threat posed by global climate change to be serious. These surveys also show that concern about climate change is associated with the perception that renewable energy is important and with the willingness to pay more for clean power, in particular wind energy. Of those surveyed, 90 percent favored the development of renewable energy alternatives to oil, and 75 percent were willing to pay more for clean power. When asked about a utility's plan to build or purchase a new generating facility to meet future customer demands, most people indicated they preferred wind turbines; in fact, supporters of wind turbines outnumbered supporters of nuclear plants by a factor greater than 4 and supporters of coal plants by a factor greater than 7.[16]

The institutionalization of the environmental movement and the relatively widespread dissemination of environmental awareness in many industrialized countries can partly account for the rapid growth of the wind energy industry in some countries. As pointed out above, the degree to which frame alignment strategies are successful depends on the ability of social movements to connect movement issues with cultural themes that arise in particular historical moments. An important energy-related cultural theme over the last three decades has been energy independence. The oil crisis that gripped the world during the 1970s and the unpredictable fluctuations in the price of imported fuels such as natural gas in subsequent years have made energy independence a priority for many politicians, at least rhetorically. Wind power is one of the public's favorite forms of energy production today not only because environmental campaigns succeeded in raising awareness of the environmental problems associated with burning fossil fuels and nuclear fission, but also because wind power struck a chord with the theme of energy independence.[17] For example, an American congressman who supported the Wind Energy Systems Act of 1980 proclaimed that voting for the act would "send OPEC a message that we are not going to continue to depend on their narcotic" and that the development of the wind power industry would help to end "dependence on OPEC's needle."[18] Having determined that the environmental movement is able to shape the growth of the wind energy industry, in the next section I examine exactly how this happens.

A Model of the Environmental Movement's Impact on the Wind Energy Industry

It is helpful to think of the growth of the wind energy industry as being influenced by the environmental movement through four pathways. Figure 1.3

presents a heuristic model—a simplified description of the agents influenced by the environmental movement, as well as the processes through which the movement influences them. Two characteristics distinguish the first three pathways from the fourth one. The first three pathways (represented by the top three rightward-pointing arrows on the left side of the model and continuing across) illustrate intended consequences of the environmental movement for the growth of the wind energy industry in developed countries, while the fourth pathway (represented by the bottommost rightward-pointing arrow) illustrates unintended consequences of the movement for the growth of the industry in developing countries. In addition, the first three pathways occur at the national and subnational levels, while the fourth pathway is present at the supranational level.

The first pathway is the influence that environmental activists, organizations, and research institutes have on energy policymakers' decision to adopt and implement pro–renewable energy policies. Environmental groups and activists influence the development of the wind energy industry by lobbying state and regional governments to adopt pro–renewable energy policies, as well as by contributing to their implementation. Many studies of social movement outcomes have shown that movements influence the adoption and implementation of policies both directly, by lobbying elected officials, and indirectly, by changing the public's preferences and concerns, but these studies have not examined the effect of policies on new industries. Many studies of industry creation have demonstrated that national industrial policies are shaped by political institutions, but they often neglect the role of social movement actors. Understanding the development of the wind energy industry requires bridging these separate bodies of literature. Thus, the model proposed here examines the direct and indirect influences of the environmental movement on the adoption and implementation of energy policies.

The second pathway is the influence that environmental groups and activists have on energy consumers by creating residential and nonresidential demand for clean energy. Environmental groups can influence organizations' decisions to switch from conventional to renewable energy. Research has shown that organizations are frequently shaped by formal and informal pressures exerted by cultural expectations in the society within which they function as well as by other organizations upon which they are dependent.[19] While most studies have focused on the role of government actors, resource suppliers, or competitors, recent studies have argued that social movements are also key actors because they can challenge common understandings and taken-for-granted assumptions about legitimate claims for organizational change. Building on these studies, the model emphasizes that the environmental movement can pressure various organizations to purchase green power through petitions, lawsuits, boycotts, and protests. The movement can also pressure organizations such as companies, universities, or local governments from inside, by changing the ideological commitment of organizational

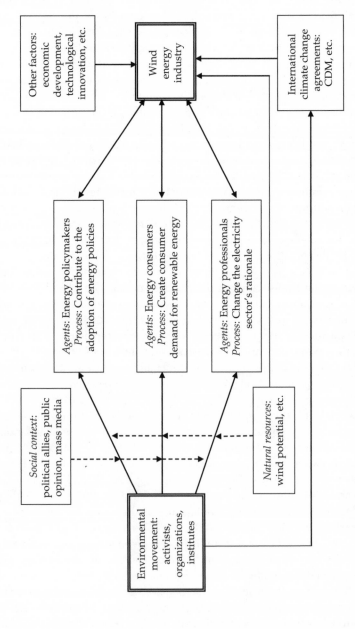

Figure 1.3. A model of the impact of the environmental movement on the wind energy industry.

members and making top executives sympathetic to the goal of environmental protection. Moreover, environmental organizations play an important role in raising awareness of the problems associated with fossil fuels and nuclear energy and in changing the public's perception of electricity production and consumption. As more and more people become concerned about the environment, and as electricity markets are deregulated and more options are available to consumers, growing numbers of residential customers decide to purchase green power.[20]

The third pathway is the influence the environmental movement has on energy professionals by changing the electricity sector's rationale. The movement can raise awareness of the environmental consequences of conventional energy production not only among electricity consumers but also among professionals working in the electricity sector. Environmental groups can push utility companies to invest in renewable energy by using external tactics such as protests and lawsuits, as well as by converting energy professionals into renewable energy champions who advocate for change from inside. Additionally, the environmental movement can stimulate entrepreneurial activity in the wind energy industry because environmental norms and cultural frameworks shape wind energy entrepreneurs' perception of social opportunities and their motivation to take risks to exploit these opportunities.[21] Studies of entrepreneurship have shown that changes in norms and cultural understandings can generate new practices in various organizational fields, and have recently begun to explore how movements contribute to these changes.[22] The model builds on these studies and highlights the role of activist-entrepreneurs and renewable energy advocates.[23] Some environmental activists interested in building an alternative electricity sector have become wind power entrepreneurs, innovators, advocates, or champions; others have founded wind-farm companies or pressured utilities to invest in renewable energy.

The fourth pathway by which the wind energy industry is affected by the environmental movement is the influence of environmental groups on international environmental agreements such as the Kyoto Protocol. Environmental organizations put pressure on national governments to join international climate change negotiations and reduce emissions of greenhouse gases. Because greenhouse gases are mixed throughout the atmosphere, what matters is the overall global reduction in greenhouse gases. Consequently, industrialized countries that ratify the Kyoto Protocol have the option to use the most cost-effective reduction in greenhouse gases and can collaborate with other countries through three mechanisms: the clean development mechanism (CDM), joint implementation, and emissions trading. These mechanisms contribute to the transfer of renewable energy technology and result frequently in the construction of new wind projects in developing countries. Unlike the previous three pathways, however, this pathway represents mostly an unintended consequence of the environmental movement. Many environmental groups are either ambivalent toward or against the creation of an international market for carbon and perceive the CDM or other similar mechanisms as governmental "greenwashing," that is, the use of a green ceremonial facade to create the false appearance of doing what is good for the environment. Also, as noted earlier, unlike the previous three pathways, which illustrate processes that operate at national and subnational levels, this pathway illustrates a supranational process.

Figure 1.3 also shows that the environmental movement's influence on the development of the wind energy industry is mediated by two exogenous factors: social context and natural resources. Social context includes political allies, public opinion, and mass media. A vibrant environmental community is an important prerequisite for a high level of support for renewable energy among policymakers and the general public. However, without strong political allies or favorable public opinion, or both, it is unlikely that environmental groups will be able to significantly shape the adoption and implementation of energy policies. Similarly, without a committed and unbiased mass media, it is unlikely that environmental groups will have much impact on the policymaking process or will contribute to a significant rise in consumer demand for clean energy. The model includes wind potential as a natural resource that mediates the effect of the environmental movement because wind is not as reliable as conventional sources of energy. Inhabitants of regions with strong environmental movements are likely to perceive wind power as a viable substitute for fossil fuels or nuclear energy, particularly if these regions also have excellent wind potential.

The environmental movement is not the only force that shapes the global development of the wind energy industry. Events like the 1973 oil crisis and nationalist movements in OPEC countries shook the utility systems in developed countries during the 1970s (Hirsh 1999; Podobnik 2006). The economic development imperative also plays an important role in the growth of the wind energy industry, particularly in countries that have little electricity in rural areas, such as India, or in regions where the opportunity to manufacture wind turbines is seen as a solution for high unemployment, as in China. Furthermore, as the wind energy industry matures, it develops its own lobby that pushes for the adoption of renewable energy policies, changes the electricity sector's rationale, and creates demand for renewable energy through "green marketing"—this is indicated in figure 1.3 by the bidirectional arrows between the wind energy industry and the agents influenced by both it and the environmental movement.

Another way to think about the environmental movement's influence is to describe the mechanisms that characterize the "dynamics of contention" over energy production and consumption. A number of scholars have recently called attention to the importance of studying the mechanisms of contentious politics.[24] Some mechanisms are cognitive, meaning that they "operate through alterations of individual and collective perceptions"; others are relational, meaning that they "alter connections among people, groups and interpersonal networks"; and still others are environmental, "meaning that they are externally generated influences on conditions affecting social life."[25]

The controversy over energy production and consumption is a form of contentious politics. According to Tilly and Tarrow (2007, 4), contentious politics "involves interactions in which actors make claims bearing on someone else's interests, leading to coordinated efforts on behalf of shared interests or programs, in which governments are involved as targets, initiators of claims, or third parties." Some forms of contention over energy production and consumption involve governments directly; for example, contention between environmental groups and governments over the adoption of renewable energy policies. Other forms involve governments indirectly as third parties; for example, contention between residential electricity consumers and utilities over the consumers' right to clean energy, or contention between wind energy developers and utilities over the right to connect to power lines.

The environmental movement influences the growth of the wind energy industry mainly through cognitive mechanisms, such as attribution of threat and opportunity and shifts in identity, or through relational mechanisms, such as brokerage and coalition formation. For example, the environmental movement can increase the perception of the threat from global climate change among policymakers, energy professionals, and energy consumers, and can create the perception of an opportunity to address this threat by the use of clean energy such as wind power. The environmental movement can also contribute to a shift in the identity of many policymakers, energy professionals, and consumers by enhancing the saliency of the "environment identity." Moreover, environmental movement leaders can act as brokers who connect environmental organizations with other movement organizations—such as labor groups—and form strong advocacy coalitions that support pro–renewable energy policies. However, as figure 1.3 shows, the effect of the environmental movement is mediated by environmental mechanisms such as wind power potential.

As a final note, it is important to emphasize that the model represents a simplified description of the environmental movement's influence on the wind energy industry. The model does not include fossil-fuel and nuclear energy industry actors, not because they do not impact the wind power industry, but because their influence stems from the social context by limiting the availability of political allies, by influencing media bias, and by swaying public opinion. Similarly, the availability of fossil-fuel resources is not included in the model because, like the level of economic development or the structure of the national power grid, it is a factor that shapes the growth of the wind energy industry independent of the environmental movement. The following section focuses on the purposive consequences of the environmental movement and uses quantitative analysis to investigate some of the model's predictions.

Quantitative Analysis: Hypotheses and Data

The theoretical model summarized above makes a number of predictions that can be tested empirically. First, the model suggests that environmental movement organizations can have a direct influence on the development of the wind energy industry. For example, environmental groups could contribute to the adoption of renewable energy policies either by direct lobbying of policymakers or by changing the public's preferences and intensity of concerns regarding energy production. They could also influence the implementation of energy-related policies by pressuring utilities to cooperate with wind power generators in order to avoid formal and informal sanctions. In addition, the environmental movement could shape the energy sector by creating consumer demand for clean, renewable energy as well as by changing energy-related values and attitudes and "converting" energy professionals into wind energy champions and entrepreneurs.

Second, the model suggests that the effect of environmental organizations is mediated by two factors: the availability of wind resources, and social context, such as the presence of political allies. The availability of high-quality wind is likely to contribute to the growth of the wind energy industry predominantly in countries where the environmental movement can create the public perception that this natural resource is as valuable as other energy resources. Similarly, the environmental movement's ability to shape energy policies and contribute to the growth of the wind power industry is likely to be greatest in countries with favorable political contexts. In fact, it is conceivable that countries with relatively strong domestic environmental groups do not adopt pro–renewable energy policies because these groups lack influential political allies or because the countries' governments respond negatively to pressure from environmental groups, either domestic or transnational. It is also conceivable that countries that have relatively weak domestic environmental groups may adopt pro–renewable energy policies if these groups have political allies.

Based on these arguments, the following orienting research questions can be formulated: (1) *Does the wind energy industry grow faster in countries with stronger environmental organizations than other countries?* (2) *Does the wind energy industry grow faster in countries with stronger environmental organizations and higher-quality wind than other countries?* and (3) *Does the wind energy industry grow faster in countries where the environmental movement has stronger domestic organizations and more influential political allies than the environmental movement in other countries?*

Regrettably, not all of the model's predictions can be tested empirically through quantitative analysis. The model also predicts that the effect of environmental organizations is mediated by the presence of favorable public opinion or a committed and unbiased mass media. The role of public opinion and mass media cannot be assessed using quantitative analysis because of difficulties inherent in data collection for studies with many cases. Currently, there are no data about public opinion and mass media related to environmental protection for cross-national datasets.[26]

It is possible, however, to use quantitative analysis to address other research questions. For example, it is plausible that various factors such as the level of economic development, the level of democratization, and the electricity sector's degree of pollution shape the development of the wind energy industry at national and subnational levels. Advanced industrial societies may develop wind energy industries because they have dense communication networks, well-trained technical experts, and cutting-edge manufacturing technologies—all of which are important prerequisites for industrial innovation and new industry entrepreneurship. Countries with a "dirty" electricity sector may develop a wind energy industry because they need to address local environmental problems such as air pollution or nuclear waste disposal. Political factors such as the level of democratization may contribute to the growth of the wind energy industry because democratic countries allow the expression of opposition to governmental policies that challenge the dominant economic priorities and well-funded lobbyists of the non–renewable energy industry. Indeed, research shows that membership in domestic and transnational environmental groups is associated with the level of democracy in a polity.[27] Therefore, an additional orienting research question is (4) *To what degree does a country's level of economic development, level of democratization, and electricity sector pollution matter for the growth of the wind energy industry?*

It is also plausible that the adoption of renewable energy feed-in tariffs (FITs) has important consequences for industry growth because they contribute to the creation of reliable markets for renewable energy producers by mandating that utilities purchase independently generated power at predetermined costs. Indeed, previous research based on case studies suggests that "reliable power purchase contracts are perhaps the single most critical requirement of a successful renewable energy project."[28] Therefore, a final orienting research question can be formulated: (5) *Are countries that have adopted renewable energy feed-in tariffs more likely to have a developed wind energy industry than countries that have not adopted renewable energy feed-in tariffs?*

This section addresses the above orienting research questions using multivariate regression analysis. This form of analysis allows examining the impact of several independent variables on the dependent variable; its role is to test the robustness of relationships between two variables while controlling for other variables. The dataset for this study is composed of data from 143 countries, representing almost 80 percent of the countries in the world.[29] The dependent variable is the level of development of each country's national wind energy industry, measured as the percentage of electricity generated from wind at the end of 2008; the variable was transformed with the natural logarithm (ln) to stabilize skew in the data. The percentage of electricity generated from wind was chosen instead of alternative measures,

Table 1.1. Means, Standard Deviations, and Sources of Variables Used in the Regression Analyses

Variable	Mean	Std. Dev.	Source and coding
Percentage of electricity from wind	0.39	1.23	World Wind Energy Association, "World Wind Energy Report 2008," and International Energy Agency; Percentage of electricity from wind
GDP per capita (ln)	8.59	1.13	World Bank; World Development Indicators for 2003–GDP in constant 1995 U.S. dollars (natural logarithm)
Democracy measure	3.20	6.40	ESI*/Polity IV Project; Democracy scores for the period 1993–2002, adjusted for trend
Coal consumed per square kilometer	1.63	3.86	ESI*/Energy Information Administration; Terajoules (TJ) coal consumed per populated land area (2003)
Nuclear energy	5.75	14.59	International Energy Agency; Percentage of electricity produced from nuclear power (2003)
Wind power potential	2.83	2.24	Archer and Jacobson 2005; Highest wind class at 80 meters (velocity > 6.9 m/s)
Feed-in tariffs (FITS)	0.19	0.39	International Energy Agency, Global Renewable Energy Policies and Measures Database; Feed-in tariffs for wind energy
Environmental movement organizations	0.59	0.95	ESI*/International Union for Conservation of Nature (IUCN); Environmental nongovernmental organizations (NGOs) per million in population (2003)
Political allies	10.12	5.69	ESI*/Union of International Associations; Number of environmental intergovernmental organizations (2003)

* *Environmental Sustainability Index* (Esty et al. 2005)

such as revenue from wind turbine manufacturing or total number of wind turbine manufacturing companies, because these measures are severely limited by missing data.[30]

Table 1.1 provides a summary of covariates and sources used in the cross-national regression analyses. The level of economic development was measured in constant 1995 U.S. dollars using information about GDP (gross domestic product) per capita in 2003 from World Bank Development Indicators.[31] This variable was transformed with the natural logarithm to stabilize skew in the data. The level of democratization was measured using information from the Environmental Sustainability Index to determine each country's average democracy score between 1993 and 2002, adjusted for trend. The electricity sector's pollution was measured using data from the Environmental Sustainability Index giving the amount of coal consumed per populated land area, and from the International Energy Agency giving the percentage of electricity that comes from nuclear power plants.[32]

The wind power resource endowment, or potential, was measured using information from Archer and Jacobson's evaluation of global wind power potential (2005). The maps of wind speed extrapolated at eighty meters—the average height of wind turbines—were used to code a variable with values from 1 (worst wind resources) to 7 (best wind resources) for each country in the dataset. The adoption of FITs was coded as a dummy variable using information from International Energy Agency, Global Renewable Energy Policies and Measures Database. The strength of environmental organizations was measured using information from the Environmental Sustainability Index, which uses information from the International Union for Conservation of Nature (IUCN). The density of environmental movement organizations was calculated by averaging the number of environmental nongovernmental organizations (NGOs) per million people.[33] The presence of political allies was measured using data from the Environmental Sustainability Index and the Union of International Associations giving the number of environmental intergovernmental organizations in each country.[34]

The arguments presented above suggest that the growth of the wind energy industry may be shaped by the strength of environmental movement organizations as well as by the interaction between this variable and the presence of political allies and the wind power potential. They also suggest that the industry's growth may be shaped by various factors such as the level of economic development and democratization, the electricity sector's pollution, and the adoption of FITs. Regression analysis is used to address the above research questions.

Quantitative Analysis: Results and Interpretation

Table 1.2 shows the correlation coefficients and the levels of significance for most of the variables included in the regression analysis.[35] The levels of

Table 1.2. Correlation Coefficients for Continuous or Ordinal Variables included in the Quantitative Analysis and Number of Cases

	1	2	3	4	5	6	7	8
1. Percentage of electricity from wind	1	.384**	298**	.133	.090	.298**	.098	.401**
		142	139	140	142	142	142	142
2. GDP per capita (ln)		1	.458**	.393**	.431**	.607**	.291**	.428**
		142	139	140	142	142	142	142
3. Democracy measure			1	.269**	.379**	.450**	.302**	.352**
			139	138	139	139	139	139
4. Coal pollution				1	.315**	.457**	-.022	.154
				140	140	140	140	140
5. Nuclear power					1	.365**	-.026	.298**
					142	142	142	142
6. High-quality wind						1	.137	.491**
						142	142	142
7. Environmental NGOs per million in population							1	.028
							142	142
8. Environmental intergovernmental organizations								1
								142

** p = .01

statistical significance refer to the probability of being wrong, that is, the probability of stating that a relationship exists between variables, when in fact it does not. The most common levels are a one-in-twenty chance of being wrong ($p = .05$), a one-in-one-hundred chance of being wrong ($p = .01$), and a one-in-one-thousand chance of being wrong in stating that a relationship exists ($p = .001$).

The variables that measure the level of economic development, the level of democratization, the wind potential, and the presence of political allies correlate positively and significantly ($p = .01$) with the variable that measures the level of development of the wind energy industry. Their association, however, is not very strong—it varies between .29 and .40. The variables that measure the electricity sector's pollution and the strength of environmental movement organizations do not correlate significantly with the variable that measures the level of development of the wind energy industry. Because correlation analysis examines the association between two variables without controlling for other variables, multivariate regression analysis is performed to assess whether or not the effects of the independent variables on the dependent variable are robust.

Before running the regression analysis, a number of multicolinearity tests were performed. These tests did not reveal any significant problems due to high correlation between independent variables.[36] The main results from multivariate regression analysis are presented in table 1.3. These results offer a negative answer for the first research question: The wind energy industry does not grow faster in countries with stronger environmental organizations than other countries. However, the results offer a positive answer for research questions 2 and 3: The wind energy industry grows faster in countries with stronger environmental organizations and higher-quality wind, as well as in countries where the environmental movement has stronger domestic organizations and more-influential political allies.

Model 1, representing the first orienting question, shows that the variable that measures the strength of environmental movement organizations does not have a significant direct effect on the variable percentage of electricity from wind. However, models 2 and 4 show that this variable has a significant mediated effect; in other words, this variable interacts significantly with the availability of high-quality winds ($p = .001$), and with the presence of political allies ($p = .01$). Countries with a high density of environmental organizations are more likely to develop wind energy industries only if they also have very good wind potential or their environmental organizations have political allies. Taken together, these results offer support for the heuristic model's assumption that the environmental movement matters for the development of the wind energy industry, but its effect is mediated by social context and natural resources.

The results in table 1.3 also offer a positive answer to the fifth research question, but mostly negative responses to the fourth research question. Countries that adopted FITs are significantly more likely ($p = .001$) to produce

Table 1.3. Estimates of the Global Development of the Wind Energy Industry at the End of 2008 (percentage of electricity produced from wind by country)

	Model 1	Model 2	Model 3	Model 4
Economic and political factors				
GDP per capita (ln)	.087*	.062	.073*	.062
	(.039)	(.037)	(.037)	(.036)
Democracy measure	.007	.005	.006	.006
	(.006)	(.005)	(.005)	(.005)
Electricity sector's pollution				
Coal consumed per square kilometer	.001	.003	.006	.005
	(.009)	(.008)	(.008)	(.008)
Percent energy from nuclear power (ln)	−.055*	−.040	−.066*	−.064*
	(.028)	(.027)	(.027)	(.026)
Wind power potential				
High-quality wind	.031	−.008	.011	.009
	(.018)	(.020)	(.018)	(.017)
Renewable energy feed-in tariff policy				
Feed-in tariffs (FITs)	.417***	.430***	.382***	.403***
	(.086)	(.083)	(.083)	(.081)
Environmental movement's impact				
Environmental movement organizations	.029	−.101	.027	−.150
	(.041)	(.052)	(.039)	(.071)

	(1)	(2)	(3)	(4)
Environmental movement organizations x high-quality wind		.069***	—	—
		(.019)		
Political allies			.021***	.011
			(.006)	(.007)
Environmental movement organizations x political allies				.021**
				(.007)
Constant	-.734*	-.461	-.756*	-.579*
	(.294)	(.290)	(.282)	(.280)
Adjusted R-square	.38	.44	.43	.46
Number of countries	137	137	137	137

*p < .05; **p < .01; ***p < .001

a higher percentage of electricity from wind. The effect is relatively large: a country with a FIT policy is likely to produce approximately 1.3 percent more electricity from wind than a country that is similar in other respects but did not adopt a FIT policy. These results should be interpreted cautiously, since the quantitative analysis does not take into account the length of time since the FITs were adopted, the specific values of the FITs, or the length of the contract terms. However, they offer additional support to the argument that FITs are the most efficient policy for stimulating the growth of renewable energy industries (Gipe 2006; Mendonça 2007). The level of economic development has a moderate and positive effect that is marginally significant (p = .05), yet the level of democratization does not have a significant effect. The percentage of nuclear power has a negative but marginally significant effect (p = .05), while the amount of coal consumed does not have a significant effect in any of the models. Therefore, countries with a high level of economic development are somewhat more likely to produce electricity from wind; also, countries that produce little or no energy from nuclear power are somewhat more likely than other countries to have a developed wind energy industry.

The most important results of the quantitative analysis suggest that while environmental movement organizations impact the development of the wind energy industry, their effect is mediated by the presence of political allies and the availability of high-quality wind. Interpretation of the quantitative analysis can be aided by briefly examining a few cases that fit the model well. For example, Germany and Denmark have strong environmental movements that have mobilized frequently against coal and nuclear energy and in support of renewable energy. Moreover, these countries have good wind resources—particularly in the coastal regions—and a social and political context mostly favorable to environmental groups during the last four decades. Consequently, Germany and Denmark were the first countries to adopt FITs and other policies that support wind power, and both countries are world leaders in the wind energy industry. The United States has a less-developed wind energy industry although it has a relatively strong environmental movement and very good wind potential. This is mostly because the social and political context in the United States has been less favorable for environmentalists. The influence of a powerful fossil-fuel lobby in America has resulted in an inconsistent federal policy to support renewable energy—although there are important regional differences.

The quantitative analysis has a number of limitations, mostly due to variable measurement issues. For example, the variable wind power potential was measured as an ordinal variable because it is extremely difficult to calculate the wind power potential in megawatts for each country. Consequently, the differences in wind potential between some countries are underestimated. Similarly, the variable density of environmental organizations does not include the number of members in each environmental group and is biased toward larger organizations. Additionally, the role of other factors identified in the model, such as public opinion or mass media, could not be assessed due to a lack of data. Finally, the quantitative analysis cannot examine the specific pathways through which environmental activists, organizations, and research institutes influence the growth of this renewable energy industry, and it cannot describe the contentious processes that shape the electricity sector's transformation. Detailed case-study analyses are necessary to overcome these limitations—the following chapters employ this method to examine the environmental movement's pathways of influence.

Conclusion

This chapter has combined insights from the literature on industry creation and social movement outcomes to identify the main factors that shape the global development of the wind power industry. It has built a model that illustrates the pathways through which environmental activists, organizations, and research institutes contribute to the growth of this industry. The model's fundamental assumption is that the environmental movement's influence is mediated by social context and natural resources. The model suggests that environmental movement organizations can influence the development of the wind energy industry because they can pressure policymakers to adopt and implement pro–renewable energy policies, create demand for renewable energy, and stimulate entrepreneurial activity in the renewable energy sector. Moreover, the model also suggests that the effect of environmental organizations is mediated by various factors such as the availability of wind resources, and the presence of political allies, favorable public opinion, and an involved and unbiased mass media.

The chapter has also tested the model using quantitative analysis. Results from regression analysis offer some support for the theoretical model and are consistent with previous studies on industry emergence and social movement outcomes. Moreover, these results advance research on industry creation and social movement outcomes by showing that the effect of the variable that measures the strength of environmental organizations is mediated by the natural resources and social context variables. The findings are also consistent with the argument advanced by many energy analysts that renewable energy FITs are essential for the rapid growth of the industry.[37] Taken together, results from the quantitative analysis support the main assumption that the environmental movement matters for the growth of the wind energy industry. More specifically, the quantitative analysis shows that the wind energy industry grows faster in countries that have strong environmental organizations, as well as high-quality wind and pro-environment political allies.

While the multivariate regression analysis offers a "big picture" of the global development of the wind energy industry, it also leaves unanswered a number of important questions. Using quantitative analysis to understand the factors that lead to the growth of this industry is similar to using a high-altitude photograph to understand the factors that lead to local geographical changes. The quantitative analysis offers a simplified account of the development of the wind energy industry because it is static and does not provide details about the specific processes that contribute to change in the electricity sector. For example, this analysis does not advance our under-standing of the specific mechanisms that underlie the environmental move-ment's impact, or the "dynamics of contention" over energy production and consumption. The next chapters examine the way in which the environ-mental movement influences the adoption and implementation of clean energy policies, the consumer demand for green power, and the electricity sector's entrepreneurship.

2

Environmental Campaigns and the Adoption and Implementation of Feed-in Tariffs

The environmental movement has always recognized the interdependence of energy and environmental policy. . . . So intimate is the association between energy and environmental quality—a link revealed again by the emerging problems of global warming and acid precipitation—that the nation's environmental agenda for the next decade will become energy policy by another name.
—Walter Rosenbaum, *Environmental Politics and Policy*
(Washington, D.C.: CQ Press, 1998, 262)

Success Stories: Feed-in Tariffs and Other Policies in Germany, Denmark, and Spain

There is little doubt that the development of the wind energy industry has been most successful in Europe, and in particular in Germany, Denmark, and Spain. Germany was already producing more energy from wind than any other country in the world at the end of the twentieth century. In 2007, German manufacturers of wind turbines held a world market share of almost 40 percent, earning about €6 billion in exports and directly employing more than thirty-eight thousand people.[1] By 2009 Germany had further consolidated its position as one of the world leaders: it had a total installed nominal capacity from wind energy of almost 24 GW, representing over 7 percent of Germany's electricity consumption.[2]

Denmark also has a very strong wind energy industry: as a percentage of its total electricity production, Denmark produces more wind power than any other country in the world, about 20 percent.[3] The country is also a major exporter of wind technology; in 2005, for example, it exported US$7.45 billion in energy technology and equipment, approximately 8 percent of total Danish exports and one-third of the total world market. One study estimated that one Danish company alone—Vestas—made 2,533 wind turbines in 2006, installing a turbine somewhere in the world every five hours (Sovacool, Lindboe, and Odgaard 2008). Similarly, Spain produces almost 11 percent of its electricity from wind, and the Spanish wind energy industry directly employed approximately twenty-one thousand people in 2008—the second-largest number of jobs in the wind energy industry in Europe, after Germany.[4] Together, Germany, Denmark, and Spain accounted for over 80 percent of annual wind energy installations in the European Union between 2000 and 2002. However, as more and more countries in the European Union developed wind projects, their share decreased to approximately 60 percent in 2007.[5]

Why are countries such as Germany, Denmark, and Spain world leaders in wind energy? The previous chapter showed that the wind energy industry has grown fast in countries where environmental movement organizations are strong and have political allies. It also suggested that one of the main pathways through which environmental movements contribute to the growth of the wind energy industry is by lobbying state and regional governments to adopt and implement pro–renewable energy policies. However, chapter 1 did not examine how environmental organizations and activists contribute to the adoption and implementation of specific energy policies.

This chapter "zooms in" on the big picture presented in chapter 1. It uses case studies to examine the way in which the environmental movement shapes the adoption and implementation of energy policies in Germany, Denmark, and Spain. The main arguments are that the adoption and implementation of renewable energy policies are contentious processes, and that environmental groups and activists are capable of influencing these processes when they mobilize large renewable energy advocacy coalitions and take advantage of favorable political contexts. The chapter focuses primarily on renewable energy feed-in tariffs (FITs), but also discusses the role of other energy policies.

Feed-in tariffs are basically "pricing laws" under which renewable energy producers are paid a set rate for their electricity depending on the technology used and on the size of their installation.[6] These tariffs are considered the single most important precondition for the rapid growth of renewable energy projects, and over thirty countries aside from Germany, Denmark, and Spain had adopted such policies by the beginning of 2008.[7] Wind energy and other renewable energy projects are often constructed and maintained not by utilities but by independent developers. The only way for independent developers to sell their power is through access to a utility's distribution grid and by obtaining contracts to sell and transfer electricity to the utility or to a third party through the grid, a process known as "wheeling." The biggest obstacle wind power developers face is obtaining a reliable long-term revenue stream because financial institutions often consider renewable energy projects to be risky. Creating reliable markets for independent power by mandating that utilities purchase all independent power at its "avoided cost" has been "the cornerstone of essentially every successful renewable energy strategy" (Redlinger, Andersen, and Morthorst 2002, 171). However, depending on the assumptions used, avoided cost calculations can vary significantly. Consequently, many wind energy projects are competitive only if they are protected by FITs—also known as power purchase agreements—that calculate sufficiently high avoided costs.

This chapter examines in detail the processes that contributed to the adoption and implementation of FITs in Germany, Denmark, and Spain, for two reasons. First, although wind energy industries in these countries have gone through ups and downs, these countries represent wind power success stories. Second, these countries adopted the first and most effective tariffs. For example, taken together these countries installed 31 GW of wind energy capacity, or 53 percent of the global total, between 1990 and 2005 with FITs (Rickerson, Sawin, and Grace 2007). Additionally, the chapter briefly compares these success stories with the stories of other European countries—France, Sweden, Austria, and Norway—that adopted FITs later and with various degrees of success.

The Environmental Movement and Renewable Energy Policies in Germany

Germany's FITs are seen by many energy experts as the world's most effective policies for the development of renewable energy technologies and, in particular, wind power. The German FITs involve fixed payments that are guaranteed for as long as twenty years but are lowered every year to encourage more efficient production of renewable energy. The first FIT was the Electricity Feed Act (StrEG), adopted in 1990; the second FIT was the Act on Granting Priority to Renewable Energy Sources (EEG), adopted in 2000 and revised in 2004. The German renewable energy FIT policies are estimated to account for over 70 percent of the electricity produced from renewable energy in 2005 and were expected to result in a reduction of over fifty-two million metric tons of carbon dioxide by 2010. These impressive results have been achieved at a relatively modest cost: in 2005 the extra cost due to the FITs and shared by all consumers was €0.0056 per kWh, or 3 percent of the average German household's electricity costs.[8]

Additionally, Germany has used renewable set-aside policies to create a separate market in which renewable energy projects compete among themselves. Germany has also created special financing agencies to provide loans for renewable energy projects. Loans with below-market interest rates are available from Deutsche Ausgleichsbank, a federal funding institution that provides favorable financing terms for projects in various areas such as environmental protection (Redlinger, Andersen, and Morthorst 2002). Since the early 1990s the German government has also used both investment subsidies, which paid wind project investors per kilowatt of rated capacity or as a percentage of total investment cost, and production subsidies, which paid project developers a significant amount of money per kilowatt-hour of electricity produced.

Why did Germany adopt FITs and other pro–renewable energy policies early on? How has Germany implemented these policies successfully? To answer these questions, it is important to examine how the environmental movement influenced the energy policymaking process well before the adoption of the first renewable energy FIT policy in 1990.

The environmental movement has influenced the restructuring of the energy sector and the growth of the wind energy industry in Germany through three main campaigns. The first were the campaigns against nuclear energy, which started in the early 1970s. The German government responded to the energy crisis in the early 1970s by increasing support for nuclear energy. For example, the Energy Program adopted in 1974 stated that nuclear power should be used "as much as possible" (Joppke 1993, 238). One state alone, Baden-Würtenburg, estimated that in order to meet a projected growth in electricity demand of 7 percent annually, it would have to construct eight nuclear power plants by 1990 (Joppke 1993, 97). Not surprisingly, between 1975 and 1985 the research and development funds allocated by the federal government to the nuclear power industry were between ten and twenty times larger than the funds allocated to wind and other renewable energy industries (Jacobsson and Lauber 2005, 130).

The federal government's plans to increase the use of nuclear energy, however, ran into unexpectedly strong opposition in 1974. The first major environmental campaigns against nuclear power were organized in that year in Wyhl. Environmental activists joined forces with local groups that were "on a quest for regional autonomy" and organized a not-in-my-backyard (NIMBY) campaign against a proposed nuclear reactor that could bring, in the words of one activist, "a cancerous growth of streets, industries, and cities [that] will destroy our beautiful land."[9] Activists gathered over ninety thousand protest signatures and, initially, followed a legal strategy. When they realized that authorities had chosen a contentious rather than consensus strategy, activists engaged in vigorous direct action. At the beginning of 1975, a spontaneous construction-site occupation grew into a permanent siege of several thousand people that lasted for almost one year. Faced with growing popular support for the local activists, the government of Baden-Würtenburg stopped construction in 1976 until further expert evidence could clarify the project's environmental impact. The following year, the Administrative Court of Freiburg permanently revoked the construction permit for the Wyhl plant, bringing the first major victory for an antinuclear campaign (Joppke 1993, 99).

Energized after the success of the antinuclear protests at Wyhl, environmental activists began organizing protest activities around the country.[10] In 1976, environmentalists as well as local farmers and leftist student groups formed a coalition to oppose the construction of a nuclear power plant at

Brokdorf in Schleswig-Holstein. When the state government secured the construction site with a police battalion, tens of thousands of protesters clashed with the police in a "civil-war like confrontation" and temporarily occupied the construction site (Wagner 1994, 273).

In 1977, antinuclear protesters clashed again with the police in the most violent confrontation ever registered in the Federal Republic of Germany— almost eight hundred demonstrators and police officers were injured in the battles near Grohnde in Lower Saxony. The violent nature of those protests, however, brought a negative image for the antinuclear campaign in the German mass media and public opinion. Because of this, one author has called the protests "the biggest defeat of the West German antinuclear movement" (Joppke 1993, 106). When antinuclear activists tried to organize a nationwide mass rally against a nearly completed reactor at Kalkar later in the same year, the police responded with a massive raid that efficiently dissolved the demonstration.

Learning from the protests at Brokdorf, Grohnde, and Kalkar, environmental activists changed their strategy and organized a large, nonviolent march in 1979. The "march to Hannover," appropriately organized to take place between March 25 and March 31, was "like the Carnival in Rio: a singing, clapping, and dancing mass of people." The march was a major success: more than one hundred thousand people joined in the final demonstration, proving that the antinuclear campaign was rejuvenated "not with helmets, but with flowers" (Joppke 1993, 112).

The issue of nuclear energy remained the centerpiece of environmental protests even after the massive demonstrations characteristic of the 1970s had peaked. For example, almost 50 percent of the environmental protests between 1988 and 1997 in Germany were directed against nuclear energy.[11] From 1996 to 1998, numerous militant protests were organized when nuclear waste was transported by rail. In 1997 the rail shipment of spent fuel rods from nuclear power stations to a temporary underground storage facility at Gorleben generated many peaceful demonstrations and the blocking of railway lines by sit-ins. Some protesters also engaged in more violent actions such as tearing out railway tracks, sabotaging overhead railway power lines, and arson attacks on signal switch-boxes at railway crossings. Officials estimated that over thirty thousand police had to be mobilized to protect this transport, and that the total security for this transport alone cost German taxpayers approximately US$60 million.[12] Considering that the protesters' real goal wasn't to prevent shipment of the waste but rather to make transporting it prohibitively expensive, the so-called anti-Castor protests were highly effective.[13]

The environmental movement's campaigns against nuclear energy contributed to the restructuring of the energy sector in Germany during the 1970s and 1980s in a number of ways. As discussed, NIMBY opposition to nuclear power plants delayed and even stopped construction at a number of sites, resulting in rising costs and making nuclear energy an increasingly unattractive option for investors. These campaigns also gradually changed public opinion from supporting nuclear energy to being against it. Opinion surveys show that the majority of Germans were in favor of nuclear energy during the 1970s; however, by the late 1980s nuclear energy opponents outnumbered proponents by more than 20 percent.[14] Moreover, the antinuclear campaigns contributed to the emergence of expert institutes that changed the nature of the nuclear energy debate. The Eco-Institute, for example, was founded in 1977 by scientists critical of nuclear power, and it soon became the leading antinuclear institution in Germany. Modeled after the Union of Concerned Scientists in the United States, the Eco-Institute aimed to provide local antinuclear groups with counter-expertise in their struggle with utilities and governments. In a move that "severely shook the nuclear establishment," the Eco-Institute used rigorous research to demonstrate that "growth and welfare were possible without petrol and uranium" (Joppke 1993, 127).

Finally, because of the growing public opposition to nuclear energy and the emergence of anti–nuclear energy expert institutes, the antinuclear campaigns gradually changed the energy priorities of government officials. During the years immediately after the 1973 energy crisis, most politicians believed that future energy crises could be prevented only by increasing production from nuclear energy.[15] But in the late 1970s, nuclear energy activists decided to form the Green Party and to run in local elections, primarily in districts affected by nuclear projects; their main demand was an instant nuclear moratorium. A number of Social Democratic Party (SPD) politicians also became increasingly critical of nuclear energy, and because their party refused to demand a nuclear moratorium, some defected to the Green Party.[16] In the period between 1979 and 1983, more and more environmental activists as well as experts from ecological institutes were accepted as interlocutors in public debate and official hearings—yet, they had no links to political actors. The situation changed in 1983 when the Greens entered the Bundestag (the German federal parliament) and demanded a national moratorium on nuclear power. By 1984 the SPD had labeled nuclear power a "transitory technology" for energy production, and by 1986, after the Chernobyl nuclear accident, the SPD had adopted the "exit from nuclear energy" as its official policy (Wagner 1994, 284). By the late 1980s even conservative parties had decided to accept nuclear power only as long as "all possible steps to abandon nuclear power in the future will be initiated as soon as possible" (Joppke 1993, 187). Consequently, in 2001 the German government passed legislation calling for the closing of all nuclear power plants by 2022.[17]

The second wave of environmental campaigns that spurred the reform of the energy sector in Germany started in the early 1980s and were directed against acid rain. The energy crisis of the early 1970s had had a serious impact on the German economy and led to major policy changes in the form of increased support not only for nuclear energy but also for hard coal. For example, in 1978 Germany decided to boost coal production from 67 to 90 million metric tons by 1990.[18] While this production goal was abandoned in 1982 due to the widening price gap between German coal and imported coal, production increased for a few years and Germany maintained its position as Western Europe's largest hard-coal producer. During the 1970s and 1980s the German government created incentives for utilities to use otherwise noncompetitive domestic hard coal—for example, utilities were paid from a fund financed by a special tax on electricity prices that varied from 3.25 percent in 1976 to 8.5 percent in 1989.[19] Yet, while the German government's decision during the late 1970s to increase coal production may have alleviated the public's concern about energy safety and independence, it also exacerbated its concern about acid rain.

In 1981 a cover story in *Der Spiegel* reported that large tracts of German forests might be dying as a result of acid rain. The environmental movement quickly seized the issue of acid rain and began campaigning against it. In 1981, the year that Greenpeace opened an office in Hamburg, Greenpeace activists climbed a tall chimney at a factory in that city in order to bring attention to the factory's contribution to acid rain. In 1982, members of another well-known German environmental organization, Robin Wood, attempted to bring attention to the fact that acid rain was destroying famous historical buildings by climbing the tower of the Michaels Church in Hamburg and unfurling a banner reading "Save the Michaels." In other cities, environmental activists also climbed chimneys of coal power plants, unfurling banners that read "Stop Acid Rain" and "The Forest Is Dead, Long Live Politics." And in 1984, a group called Struggle Against the Death of Forests organized in München what was probably the largest protest against forest death from acid rain, with over twenty-three thousand participants (Schreurs 2003, 96).

The German environmental movement's response to the government's plan to increase coal production did not stop at organizing mass protests and other highly visible events. The movement also impacted the political process by establishing the Green Party as an influential political actor; the Green Party, in turn, criticized the SPD's policy of building higher smokestacks as an ineffective "Band-Aid." During the 1983 electoral campaign, the Greens promoted an energy policy that was anticoal as well as antinuclear. The environmental movement also influenced members of the Free Democratic Party (FDP) to promote regulations to control emissions from power plants. For example, the FDP's Josef Ertl was able to secure cabinet approval for legislation requiring the use of desulfurization technology in 1981, and the FDP's Gerhart Baum "was influential in convincing the Ministry of Economics to agree to the idea of the Large Combustion Plant Regulation in 1982" (Schreurs 2003, 97).

Additionally, the environmental movement made a special effort to build alliances with trade unions. In the past, most trade unions had been either indifferent or opposed to environmental campaigns because they perceived them as a threat to job creation and economic growth. For example, in 1977 over forty thousand trade unionists participated in a demonstration "for coal and nuclear power" in Dortmund (Joppke 1993, 122). However, because of the environmental movement's strategy to win support from unions, by the early 1980s trade union representatives began to portray environmental protection as a means of stimulating employment, and numerous unions began to add environmental experts to their staffs and to educate their members about environmental problems. Consequently, in 1983 the Federation of German Unions issued a position paper on acid rain that called for stricter emission standards for large power plants and "the implementation of a government program that would secure employment while introducing measures to protect dying forests" (Schreurs 2003, 97).

The German environmental campaigns against acid rain resulted in more-stringent environmental regulation of coal power plants, the closing of many domestic coal mines, decreasing public support for tax subsidies for the coal industry, and a gradual decline in overall coal production.[20] These campaigns also accounted for significant changes in the European Union's environmental policies. In 1984 the German federal government organized an international conference on acid rain and transboundary air pollution, which was a first step toward the adoption of the Helsinki Protocol on the reduction of sulfur oxide emissions by 30 percent between 1980 and 1993. The German government pushed for an international agreement on nitrogen oxide, another by-product of burning fossil fuels and a major contributor to acid rain. The Protocol to the 1979 Convention on Long-Range Transboundary Air Pollution Concerning the Control of Emission of Nitrogen Oxides or Their Transboundary Fluxes (also known as the Sophia Protocol) was formed in 1988. Germany was one of the countries that pledged to reduce its own emissions of nitrogen oxide by 30 percent by 1998, and it pressured the European Community to adopt a directive that would introduce the German standards throughout the Community. While the German initiative at first met considerable resistance, it eventually contributed to the establishment of an EU directive on large combustion plants in 1988 that called for country-specific reductions (Schreurs 2003, 99).

The third set of environmental campaigns that contributed to the reorganization of the energy sector in Germany was directed against global climate change. The issue of climate change was first publicized in the early 1980s through the work of scientists such as Hermann Flohn. The German mass media's interest in the topic grew gradually such that by the late 1980s, the number of articles on climate change in some news sources was greater than the number of articles on acid rain or ozone depletion.[21] The growing visibility of the global climate change issue in the media and the rising public concern during the 1980s could be attributed to a number of events. In 1986 the German Physical Society published a report with the evocative title "Warning against the Impending Climate Catastrophe." This report was covered extensively in the media; a major newspaper, for example, published a story with a picture of the Cologne cathedral sinking into the ocean.

The German Meteorological Society joined the Physical Society's wake-up call in 1987, effectively putting the issue of global climate change on the public agenda in Germany (Coenen 1999, 226). The government took up the issue in 1987, when Chancellor Helmut Kohl called for national and international measures to protect the climate. The parliament responded to the growing public awareness of the issue by appointing a parliamentary Enquête Commission on Preventive Measures to Protect the Earth's Atmosphere. The Enquête Commission involved approximately fifty research institutes that produced about 150 studies and worked in close contact with the media (Coenen 1999, 228).

Numerous German environmental groups started campaigning against global climate change in the late 1980s. By that time, the German environmental movement had become one of the strongest in the world. For example, it was the largest in Europe in terms of absolute organizational membership and, arguably, the most successful in the world in terms of policy impact.[22] The German League for Nature Conservation and Environmental Protection (DNR), an umbrella organization representing over one hundred environmental nongovernmental organizations (NGOs), became involved in this issue in 1988 when it began providing information to the public about the Enquête Commission's work. In 1989, DNR became a founding member of Climate Network Europe (CNE), a European network of environmental groups working on climate change and the initiator of the global Climate Action Network (CAN; Beisheim 2005, 202). In 1992 DNR also began to coordinate and organize the German NGO Forum on Environment and Development (Forum U & E), which works—among other issues—on climate change. The main purpose of this forum is to develop position papers that can be used for informing the public, lobbying, or monitoring. In 1995 DNR organized Klimaforum '95, a working group that coordinated the activities of all environmental nonprofit organizations present at the first session of the United Nations Framework Convention on Climate Change Conference of the Parties (COP 1), organized by the United Nations Framework Convention on Climate Change (UNFCCC) in Berlin.

Through their involvement in CNE, CAN, and Forum U & E, representatives of DNR influenced both the German government's position on climate change and the international politics of climate change. For example, German environmental organizations that are members of DNR regularly present at international events and lobby the German government for significant national action to address climate change. A DNR position paper from 1996 asked for the formation of a World Council for Environment and Sustainable Development (WCESD) that would be able to decide on sanctions if international environmental law were breached, arguing that "the ruling of the WCESD should take precedence over decisions of the World Trade Organization" (Beisheim 2005, 205). Moreover, the position paper stated that "DNR demands from the Federal Government of Germany to increase the credibility of the negotiations for a greenhouse gas reduction protocol by implementing its own CO_2 reduction objective of 25 percent until 2005."

But the environmental campaigns against climate change did not involve only lobbying or disseminating information about climate science: environmental groups also engaged in direct actions aimed at increasing the public's support for renewable energy. Since the 1990s the global climate change issue has become a consolidating framework that has revitalized the German environmental movement and produced new waves of mobilizations against coal and nuclear energy. For example, in 1995 environmental groups collected over 650,000 signatures to show support for strong climate action and renewable energy; and most environmental protests in Germany during the 1990s were on the issue of energy.[23] Over the last decade, environmental groups also engaged in highly visible protest actions against fossil fuels, in particular coal. In 2004 Greenpeace activists pressured the German energy firm RWE to shift away from coal for electricity production. Over fifty Greenpeace activists chained themselves to excavators and other equipment at the site of a new coal power station in Cologne, a site they described as an "unreal moon landscape," and displayed banners reading "Coal Kills the Climate."[24] Consequently, environmental groups' intense efforts to raise awareness of the problems associated with the consumption of fossil fuels resulted in widespread positive attitudes toward renewable energy and in significant individual investments in collectively owned wind farms.[25] By 2002, individual German investors had installed as much as 4 GW of wind generating capacity, and almost one-third of all wind capacity in Germany had been built by associations of local landowners and nearby residents.[26]

A number of German environmental groups working on climate change specialized in coalition building. The Federation for the Environment and Nature Protection (BUND), one of the largest environmental organizations in Germany, formed a coalition in 1993 with the Association of Young Entrepreneurs (BJU) to support the implementation of an ecological tax reform. The coalition aimed to demonstrate that economic and ecological interests are not necessarily in conflict and that "the difference between the young generation of entrepreneurs and conservation groups existed only in the minds of politicians and established interest groups" (Beuermann and Jäger 1996, 213). Partly because of an aggressive marketing strategy and partly because of its uniqueness, the coalition attracted enormous media attention and was instrumental in focusing the national discourse on climate change and on the economic opportunities and competitive advantages offered by innovations in renewable energy.

Additionally, a growing number of environmental organizations became engaged in solution-oriented campaigning against climate change. According to one Greenpeace energy expert, the environmental movement underwent a major transformation in the early 1990s when it started to work with

experts on finding technological solutions to problems such as ozone depletion and global climate change:

> In the 1980s the environmental movement was against something. In
> the 1990s, you could still be against something, but it was also required
> that you have a solution. Greenpeace developed solutions very early.
> We developed a CFC-free refrigerator, and then we developed a very
> fuel-efficient car; those were projects with quite high investments, in
> terms of money. We have very high standards: we work with scientific
> institutions that have a very good reputation. For example, for our
> recent Energy Revolution report we worked with the German Space
> Agency. We've also been working with the renewable industries for
> over ten years because we need them to confirm our results. From my
> point of view, it makes sense to have energy demands that are not only
> ambitious but also achievable.[27]

Three environmental NGOs were particularly important for developing a
vision of the future of renewable energy in Germany as well as for coming up
with specific proposals for renewable energy policies. One was the Eco-
Institute, which was founded in 1977 with the goal of providing local antinuclear groups with scientific expertise in their antinuclear struggle. The other
two were the Solar Promotion Association (SF) and Eurosolar, founded in
1986 and 1988 respectively to campaign for political support for renewable
energy. In their attempt to demonstrate that economic growth and welfare
were possible without coal, petroleum, and uranium, these "expert-knowl-
edge" organizations developed the concept of cost-covering payment for
electricity generated from renewable sources. This concept was later applied
in various FITs at both federal and local levels in Germany (Jacobsson and
Lauber 2005, 132). Hermann Scheer, the founder of Eurosolar, argues that his
organization was responsible for moving the debate about FITs to the top of
the political agenda: "There were some campaigning organizations, which
pressed the political parties, the governments on the local level or the federal
level, to create legal frameworks. This process started with the foundation of
Eurosolar in 1988. Before this time different renewable energy associations
were mainly focused on technological development. They lacked a political
view on the matter. This has changed with Eurosolar. It politicized the
question."[28]

The Eco-Institute, SF, and Eurosolar had a major influence on the work of
the parliamentary Enquête Commission on Preventive Measures to Protect
the Earth's Atmosphere. In addition to recommending a 30 percent cut of
1987 carbon dioxide and methane emission levels by 2005, the commission
encouraged a fundamental reform of energy policy and included an electricity
FIT for renewable energy generation. The Enquête Commission's recommen-
dation was taken up by two conservative MPs from Northern Germany—
where interest in wind power was strongest—who proposed a FIT to support
wind energy in 1987.[29] Initially, the government resisted the proposal and

attempted to "buy off the dissenters" by initiating a 100 MW (later expanded to 250 MW) wind power demonstration and market-creation program.[30]

The minister of research and the minister of environment strongly supported the FIT, while the Ministry of Economic Affairs opposed it. By 1989, the number of politicians who were interested in wind power had grown so much that a coalition between Green and conservative MPs emerged. These politicians engaged in negotiations with the Ministry of Economic Affairs, which preferred voluntary concessions on the part of the utilities. Because the utilities believed that the proposed FIT would impose only negligible costs, and because they were absorbed in taking over the East German energy sector, they did not mobilize to oppose the FIT in 1990.[31] As one member of the World Wind Energy Council noted, "The parliament, actually the conservative liberal majority, adopted the feed-in law without actually knowing what would come out. They probably thought that this was something for their conservative friends from Bavaria, because that just helped them to sustain their small existing hydropower projects. The fact that it became so successful in the case of wind energy is something that probably many people did not expect."[32]

The first FIT, known as the Electricity Feed Act (StrEG), was adopted in 1990 partly because of environmental groups' campaigns against climate change. One of the intended purposes of the law was to level the playing field for electricity produced from renewable sources of energy by setting feed-in rates that took into account the external costs of fossil fuels and nuclear power. This law, together with the 100 MW and 250 MW wind demonstration programs, resulted in an exponential expansion of wind power in Germany from 20 MW of installed capacity in 1990 to over 1.1 GW of installed capacity by 1995.[33] By the mid-1990s, approximately one hundred thousand small investors and landowners had been encouraged by the Electricity Feed Act to become involved in the production of electricity from wind, and many of them formed community groups to share the initial cost.[34] Indeed, analysts of the German wind power industry note that "most wind power today results from the efforts of grassroots activists of one form or another. The REFIT [renewable energy FIT], at first in combination with the 100 MW wind program, made such practices commercially viable."[35]

The second FIT, the Renewable Energy Sources Act (EEG), was adopted in 2000 and amended in 2004. As in the case of the Electricity Feed Act, the German environmental groups' ability to influence the government's global climate change policies played a major role in its adoption.[36] German environmental groups pressured the government not only to support the Kyoto Protocol, which aims to cut emissions of greenhouse gases in industrialized countries to 5 percent below 1990 levels by 2012, but also to adopt an ambitious goal of reducing greenhouse gases by more than 20 percent. To reach this goal, Germany adopted and implemented the Renewable Energy Sources Act, which states: "The purpose of this Act is to facilitate a sustainable development of energy supply in the interest of *managing global warming and protecting the environment* and to achieve a substantial increase in the percentage contribution made by renewable energy sources to the power supply [emphasis added]."[37] This act obliged grid operators to give grid access to renewable energy plants and to purchase their electricity at premium prices. The German Wind Energy Association (BWE) emphasizes the act's importance: "Without this state-controlled minimum price, the wind energy sector would have had no chance against the billion Euro heavyweights of the coal and atomic energy industries on the cartel-organized energy market."[38]

Environmental groups involved in climate change activism have not only supported the adoption of renewable energy policies but also defended their implementation. The most notable events took place between 1996 and 1998, when the implementation of the Electricity Feed Act reached a stalemate. As the German wind energy industry began to mature, conventional electricity generators began a campaign to roll back the law because they were concerned about the uneven geographical distribution of the burden of utility payments and of excessive minimum prices for wind energy. Invoking violation of state-aid rules, in 1996 the utilities association (VDEW) filed a complaint with the Directorate General for Competition (a subdivision of the European commission that monitors fair competition). With the support of the Ministry of Economic Affairs, utilities proposed reducing the minimum price for wind from 90 percent of average sales price to 75 percent, as well as limiting the support mechanism in time.[39] The German utilities' opposition to the FIT contributed to insecurity for investors and stagnation in the market for wind turbine manufacturing between 1996 and 1998.

While the implementation of the FIT was delayed for a brief period by the opposition from the Ministry of Economic Affairs and the supraregional utilities, the environmental movement's mobilization prevented it from being completely derailed. In 1997 the government's proposal to reduce FITs led to a large-scale demonstration, bringing together numerous environmental groups as well as other civic groups. A call to defend the FIT, launched by the German Wind Energy Association, was answered by environmental groups such as Greenpeace and BUND, as well as by the metalworkers' union, farm groups, and church groups. More than four thousand people participated in a protest march in Bonn. One member of the World Wind Energy Association emphasizes the importance of the fact that environmental groups formed a broad coalition:

> In the mid-90s it became more and more obvious that wind energy would be a main driver and that, of course, was when the utilities started their attacks on the feed-in law. The minister for economics at that time was from the Liberal Party, the only party that's still against the feed-in tariff in Germany. He tried then to reduce the tariff and that was when there was a broad coalition formed: there were trade unions, the emerging wind industry. Religious groups were involved, farmers were involved, and all environmentalists were involved. It was a quite

broad coalition, covering a broad political spectrum. That made it quite successful and influential.[40]

At the same time, environmental groups held press conferences and distributed information about the FIT's benefits for the environment as well as the economy. For example, Greenpeace started a campaign in which it appealed to all German electricity consumers to use the new freedom offered by the liberalized electricity market and "demand green power." Greenpeace also defended wind power producers from the utilities' attack and argued that generators of clean power had to pay significantly more for transmission of their power than the utilities paid for use of their grids.[41] Greenpeace had the power to shape public perception of energy policies because it is a respected and trusted organization that publishes high-quality, solution-oriented research, such as the "Energy Revolution" report. One Greenpeace organizer offers the following explanation:

> A recent survey about the most credible institutions in Germany listed Greenpeace as number one, ahead of the police, the army, the church, or the politicians. We had about 60 percent credibility, while the politicians had about 5 percent.... So, a politician can claim that we can have 100 percent renewables in ten years; everybody loves to write about this in the media. But, a politician can do that, Greenpeace can't. While nobody really takes seriously what a politician says, when institutions say something, they have to back it up with facts.[42]

The German FIT was also attacked at the beginning of the twenty-first century. In 2001 the economics minister, Werner Müller, launched an outright attack against the Renewable Energy Sources Act. Next year Müller's successor, Wolfgang Clement, stepped up that fight and called for an immediate 15 percent decrease in the FIT and then an annual decrease of 5 percent, although the Ministry of the Environment wanted to decrease the feed-in rates gradually, by 1.5 percent per year. Large utilities joined the fight against the Renewable Energy Sources Act, arguing that they would have to use up to 7 percent of energy produced as "buffer energy" to cover the short-term variability of wind (Michaelowa 2005).

These attacks were unsuccessful, however, for a number of reasons. One reason is that environmental groups again organized a large coalition to defend the FITs and support the wind energy industry. In 2003 a large demonstration of several thousand people was organized in Berlin by more than thirty associations, including many environmental organizations such as Greenpeace, BUND, Robin Wood, and the World Wildlife Fund.[43] As one wind energy analyst noted, the support the German wind energy industry received from environmental NGOs was very important: "Almost all environmental organizations are supporting the feed-in system like it is. Just that they are publicly in favor of wind energy is, of course, important. So, it is definitely helpful that the major environmental NGOs are all supportive of wind energy."[44] Another reason for the unsuccessful attacks is that environmental groups had influential political allies not only in the Green Party and the SPD, but also in the Christian Democratic Union (CDU). Indeed, one CDU member who declared his support for the wind energy industry stated that "in this matter we collaborate with both the Greens and the Communists."[45]

Still another reason is that the wind energy industry had grown so big that its proponents could claim that it played an essential role in Germany's economy. Representatives of the German wind energy industry stressed that by 2002 the industry had created thirty-five thousand jobs and was adding three thousand more each year, particularly in economically weak regions in East Germany. Moreover, they emphasized that wind power manufacturers had become the second most important customer for the German steel industry, after automobile companies. Consider the following interpretation offered by one World Wind Energy Association analyst:

> I would say the environmental awareness is an important catalyst, but also the fact that farmers are earning money out of wind energy. That's a new sort of income for them. This kind of distributed wealth and additional wealth generation has been quite important. This is similar in Eastern Germany. Take the example of Enercon in Magdeburg; they had, I think, a train wagon factory there that employed twenty thousand people and all of them were fired after the unification. Today, Enercon employs three thousand or four thousand people there. It's of course fewer people than in the past, but it's much better than nothing and people see that Enercon is the biggest industrial employer today in Magdeburg and in the state of Saxon-Anhalt.[46]

Finally, it is important to emphasize that the climate change campaigns have been successful in Germany not only because environmental groups mobilized effectively but also because they had influential political allies. Indeed, some of the governing parties such as the Green Party and the SPD have basically adopted the environmental movement's concerns.[47] The German government has also created numerous environmental agencies and institutes. For example, in the late 1980s a new Federal Ministry for Environment, Nature Protection, and Nuclear Safety (BMU) was established, which centralized formerly dispersed competencies for environmental protection. In 1990 the German Federal Environment Foundation was formed with the aim of funding environmental protection programs; in 1991 the Wuppertal Institute for Climate, Environment, and Energy was created with the aim of promoting energy saving and renewable energy; and in 1992 the German Advisory Council on Global Climate Change was set up to develop climate change policies.[48] Environmental organizations routinely participate in hearings and conversations with ministries, and many representatives of the Federal Ministry for Environment praise the work of these organizations. And some environmental groups are funded by federal and state governments "with the declared objective to create a counter-lobby" (Brand 1999,

52). Not surprisingly, Germany is considered one of the greenest nations in the world.[49]

Furthermore, climate change campaigns in Germany have also been successful because the German mass media has been involved and unbiased and public opinion has been favorable to strong environmental action. The German mass media has presented the global climate change issue mostly as a major threat to humanity and as an opportunity to enhance German leadership on renewable energy technology. Environmental protests and campaigns receive positive coverage in the mass media, and environmental groups such as Greenpeace are respected and trusted sources of information. The vigorous mobilizations by environmental organizations, combined with the presence of a favorable political context and an involved and unbiased mass media, have contributed to the formation of a public opinion that is sympathetic toward environmental issues and supportive of FITs and other renewable energy policies. Herman Scheer argues that this combination of factors has resulted in a cultural shift in the perception of energy:

> It is no doubt that Germany has made more progress in the implementations of RE [renewable energy] than other countries. It is the result of campaign activities... Campaign activities mean to create an offensive, to speak about all the benefits of RE in order to create a public consciousness on RES [renewable energy sources]. Then people demand political steps from their own government. Media play an important role. Independent associations like renewable energy associations, environmental organisations, etc. play an important role in creating such consciousness. We speak about a cultural process here.[50]

The Environmental Movement and Renewable Energy Policies in Denmark

Denmark is another European country that has adopted a highly successful FIT. Between 1979 and 1992 the Danish wind electricity market was regulated by voluntary agreements between the utilities and associations of wind turbine manufacturers and owners. These power purchase agreements had two goals: to make the utilities purchase electricity from wind turbine owners at a guaranteed minimum price, and to distribute between utilities and turbine owners the cost of connecting wind turbines to the power grid. The agreements were designed to encourage small-scale wind turbine development; initially they applied only to individually owned turbines of less than 150 kW and to cooperatively owned turbines located near the owning cooperative. In 1992 however, the voluntary agreements broke down when the utilities, manufacturers, and owners disagreed over the price of wind power per kilowatt hour and the distribution of grid-strengthening costs. To resolve the stalemate, the Ministry of Energy and Commerce adopted a FIT that maintained the previous payments for wind power—85 percent of normal residential prices—and stated that wind turbine owners had to pay for connecting their turbines to the low-voltage grid while utilities had to pay for strengthening the high-voltage grid (K. Nielsen 2005, 116). The tariff was terminated in 2000 as a result of electricity liberalization; instead of the 85 percent price, a fixed, nominal payment was enforced.[51]

Denmark adopted a number of other policies that stimulated the growth of the wind energy industry. In 1984, the government introduced a general per-kilowatt electricity tax on electricity consumption that exempted renewable energy. In the same year it introduced a per-kilowatt subsidy for producers of electricity, and, starting in 1991, an additional "CO_2 subsidy" for all producers of renewable energy and a subsidy for wind power producers. In 1985 the Danish government introduced its first renewable set-aside (or quota), which committed utilities to install 100 MW of wind power before 1990. In 1996 the government required utilities to install another 200 MW through another renewable set-aside policy, and in 1998 it required another 750 MW from offshore wind farms. From 1979 to 1989 the Danish government also provided direct capital investment subsidies for wind turbine projects—it is estimated that more than 2,500 wind turbines were subsidized with over kr 270 million.[52]

It is not possible to fully understand why Denmark adopted and implemented the various policies that made it one of the global leaders in wind energy without examining the role of the Danish environmental movement. Similar to its German counterpart, the Danish environmental movement contributed to the growth of the wind energy industry through different campaigns. One of those was the campaign against nuclear energy. The energy crisis of the early 1970s had a severe impact on the Danish economy because imported oil accounted for 88 percent of Denmark's total energy supply—a significantly larger percentage than in other Northern European countries.[53] The Danish government responded to this crisis by increasing its support for nuclear energy. Although Denmark had no nuclear power plants at the time, at the end of 1973 the government announced that it was speeding up nuclear development, and one utility published a list of ten potential sites for nuclear power plants.

The Danish environmental movement at that time was dominated by two main groups: the Danish Conservation Society, an association founded at the beginning of the twentieth century whose mission was to conserve nature for the people; and NOAH, a group rooted in the alternative political culture of the 1960s. Following the government's announcement of its plan to develop nuclear energy, a new group was formed by people involved in NOAH and other environmental associations. This group was named the Organization for Information about Nuclear Power (OOA). Because OOA had a minimal program, an open structure, strict political neutrality, and a simple goal (to stop nuclear power), it organized a highly effective campaign in a very short time. Two planks made up OOA's strategy: demanding that the decision to build nuclear power plants be taken away from the central administration and handed over to the parliament, and asking for more public debate and information. This strategy was so successful that the government repeatedly postponed its plans for nuclear power for fear of a referendum (Jamison et al. 1990, 98).

By 1980 OOA had quickly developed into a national grassroots movement that included approximately 130 local groups. While these groups were autonomous and operated according to the principles of direct democracy, OOA also had a national secretariat that could take initiatives and make statements on behalf of the entire organization without consulting the local groups. Because of its unique organizational structure, OOA was able to counteract powerful adversaries that were centrally organized, such as the state and energy companies (H. Nielsen 2006, 215). The OOA activists also did their best to stay ahead of nuclear power advocates with regard to objective argumentation and knowledge. In addition to organizing major demonstrations and marches such as that against the Swedish nuclear power plant in Barsebäck, they collected signatures, organized meetings, and distributed campaign newspapers. In 1979 and 1980, for example, OOA organized a major campaign in which it distributed the informational brochure "Denmark without Nuclear Power" to all households in the country.[54] Consequently, as some authors note, "no one in Denmark would doubt the crucial role of OOA: it was this grassroots organization which countered the strong political-economic interest in nuclear power and led to a Danish renunciation of nuclear power" (Jamison et al. 1990, 98).

Another organization that was instrumental in the antinuclear campaign was the Organization for Renewable Energy (OVE). Formed in 1975 mostly by OOA activists, OVE concentrated on the development of practical alternatives to nuclear power; and while somewhat smaller than OOA, it quickly gained a substantial number of members around the country. Environmental activists involved in OVE were dedicated to building an organization that was at once against the existing energy policy and for an energy policy directed toward an "alternative future."[55] What made them unique among environmental activists was their commitment to demonstrating that wind and other sources of renewable energy were feasible alternatives to nuclear power and fossil fuels. This is how one OVE founding member recollects the founding of the organization: "In 1975 OOA had a nationwide meeting at Bryrup. At this meeting it was decided to break up the organization in two parts to enable some of us to work more on the new alternatives. This was the birth of a new organization: OVE. The rest of OOA would still concentrate on gathering information on the threats of nuclear power. I still remember the last words of one of the speakers to us: 'You can turn the wheels of history'; and we all wanted to do that."[56]

The Organization for Renewable Energy had an essential role not only in inhibiting the emergence of the nuclear industry in Denmark, but also in initiating the development of the wind energy industry. It contributed to the growth of the Danish wind energy industry by pressuring legislators both to pass legislation against the construction of nuclear power plants and to adopt pro–renewable energy policies. After its formation, OVE lobbied and pushed to be represented in energy-related law-and-regulation formulating processes through hearings and committees. It also published reports that aimed to show that Denmark could solve its energy problems through conservation and renewable energy development. In 1976, for example, OVE and OOA published an alternative power plan, advocating dropping nuclear power and emphasizing the opportunities presented by energy conservation and a decentralized system of renewable energy production. By involving university scientists, energy amateurs, and professionals in a national debate about the energy plan, and by disseminating information to the general public, these organizations created a "public sphere for alternative technology exchange" and helped turn the majority of the public against nuclear power and toward renewable energy (Jamison et al. 1990, 105). In fact, a majority of the Danish public was against nuclear power by the late 1970s, and by the late 1980s nuclear power opponents outnumbered proponents by almost 50 percent (H. Nielsen 2006, 216; Kolb 2007, 211). As a result of OVE and OOA's activities, the government passed a law against the construction of nuclear power plants in 1988 and adopted a variety of policies that supported wind power and other renewable energy industries.[57]

The Organization for Renewable Energy also had a decisive impact on the growth of the Danish wind energy industry by using a practical approach and creating "counter-pictures, emotional symbols of alternative societal visions" (Jamison et al. 1990, 104). The best example is that of the wind turbine constructed at Tvind in 1978. This turbine was considered by its creators to be a "fifty-three-meter-high argument," and it became a public attraction not only because of its size but also because it was built by Tvind locals, without much help from the experts. The decision to construct a large wind turbine emerged out of the desire to prove to the world not only that wind power is feasible but also that wind is a more democratic and gentle form of energy production than nuclear power. According to those involved in building the turbine,

At Tvind people were against nuclear power, with its problems of nuclear waste and monopolization. Wind energy was common sense. There is lots of wind in Western Jutland. The wind cannot be monopolized—it blows on the poor as well as on the rich—and there are no dangerous waste products. So the idea was formed and turned into a decision to build a windmill. Tvindkraft had to be big, a proof of wind power being a real alternative to nuclear power.[58]

The turbine was a major technological accomplishment of the OVE activists involved in the folk high-schools established in the early 1970s on the

Jutland Peninsula. It took them three years to build the turbine, and they had to overcome many technical difficulties along the way. For example, because there were no companies in the world at the time that produced blades as long as twenty-seven meters—the required dimensions for the Tvind turbine—the wind energy pioneers had to learn how to work with fiberglass by building three fishing boats first, and then build the blades themselves.[59] Here is how the wind power enthusiasts described their work:

> The Mill Team was the implementing force and consisted of members of Tvind's teacher group and a long row of volunteers who by their labour wished to participate in this vigorous demonstration in the energy debate.... The blades, which no manufacturer dared commit themselves to producing, aerodynamics, calculations of strength, as well as practical execution had to be developed from scratch. The shaft, gearbox, and generator were bought second-hand, and the frequency converter control box was put together by Professor Ulrich Krabbe from Denmark's Technical University and his students.... Computer control and supervision systems were developed, and long assembly language programs were written for the Z80-computer. And finally, the large cranes arrived and hoisted all the parts up. First the cap, and then the generator and gear box on top, the main shaft with the hub, and finally the blades, one at a time. Like that.[60]

The construction of one of the largest wind turbines in the world by a group of wind power enthusiasts with little technical experience convinced many Danes that the answer to the country's energy problems was literally "blowing in the wind." Their determination became a source of inspiration for other people. One wind power enthusiast describes the effect of the Tvind wind turbine: "A lot of people volunteered to build the turbine. The building of this big machine gave confidence to people: We do not have to wait for governments or big companies, we can do it together. In the following years many self-builders visited Tvind, picked up inspiration and built their own smaller turbines.... This turbine was a beacon and a great inspiration for the early Danish wind power community."[61] Another wind power enthusiast describes the impact of the turbine this way: "The effect of the Tvind mill as a source of inspiration cannot be overstated. A large number of the pioneers became hooked, like me, on the possibilities and practical challenges of wind power when they visited Tvind. The almost nonchalant self-confidence with which the Mill Team built something no one had ever done before was very contagious."[62]

The environmental activists involved in constructing the Tvind wind turbine stated that their goal was "to show the way forward for wind energy— and to show the way out from nuclear power."[63] By most measures, they succeeded. Their actions contributed not only to the eventual abandonment of government plans to build nuclear power plants but also to a rising tide of public interest in wind power. Every weekend, Tvind was filled with visitors who wanted to learn about the impressive wind turbine. It is estimated that in 1977, when the construction was about to finish, seventy-seven thousand people passed through Tvind in a two-month period.[64] The Tvind high school established an energy office for visitors who wanted advice about renewable energy systems and published a book about the town's experience under the title *Let a Hundred Windmills Flourish*. Soon similar energy offices emerged all around Denmark. In 1977 OVE established the Cooperative Energy Offices, a national network where ordinary people could get free information and advice about renewable energy systems. Also, OVE frequently organized information sessions where members could meet and discuss common problems. Indeed, one author notes, "where OVE conferences and seminars were a central tool in knowledge production, the energy offices were the central vehicle for knowledge dissemination" (Jamison et al. 1990, 105).

Throughout its existence, OVE has maintained a dual strategy and a long-term perspective. On the one hand, OVE has persisted in its plan to influence the energy-related policymaking process by actively participating in hearings and committees, and by publishing policy papers on renewable energy. In OVE's most recent mission statement, it announces that the organization "works actively towards a goal where the supply of energy in Denmark before 2030 is based on 100 percent sustainable energy sources."[65] On the other hand, OVE has remained true to its origins as a grassroots organization even while becoming professionalized. Between 1978, when it became a regular membership organization, and 2008, OVE grew to about 3,200 members, consisting of both individuals and groups such as windmill cooperatives. Throughout this time, it continued to organize local seminars and meetings at which technicians and users shared knowledge and experiences. It also published a bimonthly magazine called "Renewable Energy and Environment," promoted information campaigns in cooperation with the government-funded Offices of Energy Services and Local Energy and Environmental Offices, and organized educational programs such as the Schools Energy Forum and the Climate Caravan.[66]

In the late 1980s, OVE and other environmental organizations started organizing campaigns against global climate change. By that time the environmental movement in Denmark had become a powerful force. For example, it is estimated that the environmental organizations during the 1980s had more members than all of the political parties combined, and that almost 15 percent of the adult Danish population in 1988 belonged to one organization alone—the Danish Society for Nature Conservation (DN; Jamison et al. 1990, 112–13). In 1992 OVE joined other organizations such as BirdLife, DN, the Ecological Council, Greenpeace, Nature and Youth, and the World Wildlife Fund and formed the Danish 92 Gruppen, a coalition of Danish NGOs working on issues related to energy, environment, and development. As a result, the general public maintained a positive view of wind power even as the density of onshore wind turbines greatly increased.[67] Furthermore, over one hundred thousand households in Denmark owned shares of wind turbines located either in their community or nearby.[68]

In 2002 OVE started a climate campaign in cooperation with Greenpeace, the World Wildlife Fund, Nature and Youth, and the Ecological Council. As part of this campaign, the NGOs published a joint report assessing whether Denmark was on track toward meeting its obligations under the Kyoto Protocol. The report criticized the Liberal-Conservative government that had come to power in 2001 for carrying out vast cutbacks in subsidies to develop renewable energy. For example, it noted that while in the past the Energy Ministry had been merged with the Ministry of the Environment because of concerns about local air pollution and greenhouse gas emissions, the new government had moved energy policy to the Ministry of Economics and Industry, "signaling a less environmentally friendly and a more 'old industry' friendly policy than previous governments."[69] The report also emphasized that "the basic reason why wind energy has such a prominent place in Danish energy planning is the need to reduce greenhouse gas emissions. Denmark had a target of reducing CO_2 emissions by 22 percent between 1988 and 2005. More than one third of that target was being met using wind energy to replace coal-fired power generation."

Finally, it is important to emphasize that, as in the case of Germany, the environmental movement's ability to influence the energy policymaking process resulted not only from environmental groups' effective campaigning but also from the presence of a relatively favorable political context, an involved and unbiased media, and positive public opinion. The grassroots mobilizations for renewable energy during the 1970s were accompanied by a flurry of media activity to educate the public about wind power. The cooperative ownership structure of many wind projects and the wide dissemination of information about wind turbines contributed to the fact that Danish public opinion strongly supported wind farms.[70] Moreover, the Danish state began to institutionalize environmental concern starting in the early 1970s. The government established a Ministry for Environmental Protection in 1971 and passed legislation in 1973 that was supported not only by Social Democrats but also by the Conservative Party.[71] The state administration gradually increased its environmental protection activities and began to employ a large staff of legal advisers and technical experts to develop more coherent environmental policies. While in the earlier period the Danish government only reacted to the environmental movement, starting in the 1970s the government's tactics became more proactive, and it started to define environmental problems, develop solutions, and make long-term plans (Jamison et al. 1990, 95).

Although the Danish Green Party was not a significant political force, environmental activists had many allies and sympathizers in the Social Democrat Party and the Socialist Folkparty. Environmental groups such as OVE or the Folkecenter for Renewable Energy received significant financial funding from various governmental agencies. For example, the Steering

Committee for Renewable Energy, which was formed in 1982 by the Danish Board of Technology, gave priority to wind energy research and directed approximately one-tenth of its funding to the Folkecenter for Renewable Energy for its so-called Blacksmith Mill design project. As the wind energy industry matured and became more competitive, the government decreased its support for renewable energy: the Steering Committee for Renewable Energy was abolished in the early 1990s, and funding for the national wind energy research program received significant cutbacks after 2001.[72] However, even after this date the government still funded regional development activities and still promoted renewable energy research and development with tax deductions. In 2007 the government established a new Ministry for Climate and Energy to "strengthen efforts against climate change and to prepare for the climate summit COP 15 in Copenhagen in 2009."[73]

The importance of an antinuclear public opinion and political allies is also emphasized by one of the most influential Danish wind power pioneers, Henrik Stiesdal. As he argues,

> We do not have the two-party system of government in Denmark, we have a multi-party system. There was always over the years a left-wing and a right-wing; but in between the two there was a fairly small party—the Radical Left Party—which would sometimes be in the government of a right-wing group and sometimes in the government of a left-wing group. They were basically always there, in one constellation or another. This small party was in the crucial years a very strong proponent of alternatives to nuclear [power], and that's what really made the government be much more active here than it would otherwise have been. So, the wind industry was tremendously helped by the fact there was a public concern about nuclear power. But it was also helped by the fact that many people were engaged in the energy movement, and this small party had some shared roots with the energy movement. The left-wing parties generally always loved wind because it was an alternative to nuclear, and they didn't like nuclear. The right-wing parties wanted nuclear; but if they had the government, they were still somehow forced to support wind because this small party had an influence.[74]

The Environmental Movement and Renewable Energy Policies in Spain

The Spanish wind energy industry took off in 1997, the year when the Electric Power Act was adopted. This law differentiates between the average rate of electricity production and what the law labels the Special Scheme for facilities using renewable energy. Royal Decree 2818—adopted in 1998—regulated producers' right to incorporate the whole of the electric power produced into the electric grid and their entitlement to be paid the price on the wholesale market plus a premium. Royal Decree 436 from 2004 modifies the legal and economic framework for electricity generation under the Special Scheme and establishes a system to support electricity generation based on the free choice of producers, who can decide between a regulated tariff and sale on the open market.[75]

In addition to these FITs, the Spanish government also adopted renewable set-aside policies that aimed to greatly increase the use of green power such that by 2010 Spain would meet 30 percent of its total electricity consumption through renewable energy sources, with more than 20 GW coming from wind. To meet its national target, Spain also created preferential financing agencies such as the Official Credit Institute (ICO) and the Institute for Diversification and Energy Saving (IDAE), which offer investment assistance to renewable energy projects through low-interest loans.

The adoption and implementation of FITs and other renewable energy policies in Spain was influenced by the environmental movement's campaigns. The campaign against nuclear energy, which started in the early 1970s, played an important part in this process. Local opposition to nuclear power was precipitated by the government's plan to expand the use of nuclear energy after the 1973 oil crisis. When the government announced plans to build a number of nuclear power plants in a highly populated area on the Basque Coast, local opposition began to form around established networks of neighborhood organizations and fishing collectivities. The death of Franco and the ensuing democratization process allowed the broadening of popular opposition, particularly against a proposed nuclear power plant in Lemoniz. In 1976 an antinuclear demonstration was attended by about fifty thousand people. In 1977 approximately two hundred thousand people marched through Bilbao to protest against the construction of the nuclear power plant in Lemoniz—the biggest antinuclear demonstration in Spain (Rüdig 1990, 138). The protests were successful partly because, while the official organizer of these demonstrations was the Commission of Defense of the Non-Nuclear Basque Coast, the main mobilization was carried out by nationalist groups associated with the Basque separatist group ETA.[76]

Because some local groups opposing nuclear power were associated with regionalist protests against the Spanish central government, the antinuclear protests frequently escalated into violence. In 1978 ETA exploded ten bombs at various sites owned by Iberduero, the utility responsible for the construction of the Lemoniz power plant. One of these bombs killed two workers and injured several others. The following year, a young woman at an antinuclear sit-in was shot dead by the authorities—this led to more violent protests and the placement of another bomb by ETA, which killed an Iberduero worker. Consequently, the town council of Lemoniz ordered construction to stop; the actual construction was stopped only in 1983, however, when the Socialist Party came to power (Rüdig 1990, 213). Elsewhere in Spain, in places where antinuclear opposition was not associated with nationalist groups, local opposition was nonviolent but also less sustained.

The Coordination of Local Councils for a Nuclear Moratorium was formed in 1979, and by 1980, most nuclear sites experienced some opposition from small ecological groups. However, throughout the 1970s the antinuclear protests remained rooted in strong regional—Basque and Catalan—identities (Rüdig 1990, 216).

Starting in the early 1980s, opposition to nuclear power turned to protests against nuclear waste disposal. In 1987, for example, almost twenty thousand people participated in protests against a nuclear waste disposal site near Salamanca. In 1989, after a fire at a nuclear power plant in Vandellos, almost one hundred thousand people demonstrated in Barcelona and demanded a phasing-out of nuclear power.[77] The antinuclear protests in Spain have contributed to a de facto moratorium on nuclear power since 1984, when the newly elected Socialist Party stopped construction at several nuclear sites. Indeed, while Spain did not have a strong movement against nuclear power at the federal level, vigorous antinuclear protests at the local level contributed to a significant decrease in the popularity of nuclear energy nationwide. It is important to note that a majority of the population in Spain was against nuclear power even before the Chernobyl nuclear accident, in contrast to countries such as the United Kingdom, France, Italy, Sweden, and Switzerland (Kolb 2007, 211).

By the early 1990s, however, Spanish environmental organizations were campaigning not only against nuclear power but also in favor of renewable energy. Using a "carrot and stick" approach, organizations such as Greenpeace constantly praised politicians who supported renewable energy policies and criticized politicians who supported pro-nuclear policies. For example, in 2006 Greenpeace and other environmental groups praised the Zapatero government for showing "true leadership" in preparing the phase-out of nuclear power. At the same time, Greenpeace was part of a national coalition of environmental and civic groups that petitioned the government to "deliver on election promises of safer, cleaner, cheaper energy."[78]

Antinuclear groups in Spain were able to influence the adoption and implementation of FITs because they organized both in opposition to nuclear power and in support of renewable energy. Local opposition to nuclear power plants delayed construction at a number of sites, resulting in rising costs and making nuclear energy an unattractive option for Spanish investors. The antinuclear protests also gradually turned the tide of public opinion against nuclear energy; in turn, the growing unpopularity of nuclear energy contributed to a gradual change in energy priorities for government officials. However, in contrast to antinuclear opposition in Germany or Denmark, the antinuclear opposition in Spain was somewhat less influential because it formed later and did not result in the emergence of "expert knowledge" eco-institutes.

The environmental movement also contributed to Spain's adoption and implementation of FITs through its climate change campaign. In contrast to other European countries such as Germany or Denmark, Spain during the

1970s had weak environmental groups that were unable to organize large-scale mobilizations or effective lobbying at the federal level. Because Spain had a dictatorial regime until 1975, the environmental movement had a lower rate of cooperation with other "new" social movements than the rate of cooperation experienced by environmental movements in other European countries.[79] However, this had begun to change by the early 1980s. Professional environmental organizations in Spain have since experienced significant growth in membership, and the number of environmental non-profits has increased dramatically. For example, Greenpeace Spain grew from approximately sixteen thousand members in 1984 to over eighty-five thousand in 2004; nationwide, the number of environmental organizations almost doubled from the late 1980s to the late 1990s.[80]

An important characteristic of the environmental movement in Spain since the early 1990s is that it has increased its mobilizing capacity even as it has become more institutionalized. On the one hand, Spanish environmental groups have mobilized growing numbers of people. For example, the percentage of the population who had taken part in an environmental demonstration increased from 6 percent in 1993 to almost 15 percent in 2004, one of the highest numbers in Europe and even in the world (Jiménez 2007, 370). On the other hand, Spanish environmental organizations have consolidated their organizational infrastructure and have gained increased access to policymaking. Numerous environmental groups joined forces in the late 1990s and formed Ecologists in Action, an umbrella organization that grew to approximately three hundred groups by 2005 (Jiménez 2007, 375). As the movement grew, the Spanish government started to include environmental groups in the decision-making process. Under the influence of the European Union, in 1994 the Ministry of Public Works, Transport, and the Environment created the National Advisory Committee for the Environment (CAMA) to allow environmental groups to participate in the decision-making process.[81] And, recognizing the increasing political significance of environmental problems, the Spanish government created the Ministry of the Environment in 1996.

But the Spanish environmental movement has done more than become professionalized and increase its capacity for mobilization; it has also changed the focus of its campaigns. Early campaigns were concentrated on local issues such as the siting of nuclear power plants or local air and water pollution. During the 1990s growing numbers of environmental groups started to campaign against global climate change. Environmental groups educated the public about the threat of climate change (for example, by emphasizing the dire consequences for Spain's water resources), protested against dirty energy, and affirmed their vigorous support for wind power and other forms of renewable energy. Surveys of Spanish environmental organizations conducted during the 1990s reveal that the development of clean energy was at the top of the agenda for most groups.[82] An analysis of environmental protests in Spain during the 1990s also shows that energy had become one of the most important issues for collective mobilization: it was second only to the problem of urban and industrial pollution, and ahead of other issues that had been the traditional focus of environmental protests, such as nature conservation or water pollution (Jiménez 2003, 191).

Greenpeace has been the most active national environmental organization in Spain over the last two decades. The organization participated in more protests during the period 1988–1997 than either of the two main Spanish coalition organizations, the Coordinating Committee of Environmental Defense (CODA) and the Ecologist Association for the Defense of Nature (AEDENAT), despite the fact that these organizations were older and included more local groups than Greenpeace Spain.[83] Together with other environmental organizations, Greenpeace has run a climate change campaign that affected the Spanish government's decision to adopt and implement pro–renewable energy policies in a number of ways.

First, Greenpeace and other environmental groups published reports that legitimated FITs and other pro–renewable energy policies by publicizing the environmental and economic benefits of wind power. A 2005 Greenpeace report estimated that wind energy could account for an impressive 11 percent reduction in emissions of greenhouse gases in the Spanish energy sector by 2011, and that the wind industry provided employment to over seventeen thousand people in Spain, with a projected rate of increase of over 10 percent per year.[84] Another Greenpeace study argues that wind generates five times the number of jobs and more than two times the power for the same investment as nuclear energy.[85] Still another report, titled "100% Renewable: An Electrical System for the Spanish Peninsular and its Economic Viability," estimates that by 2050 Spain could be producing 100 percent of its energy needs exclusively from non-fossil-fuel sources without major economic costs.[86]

A particularly influential study commissioned by Greenpeace had the title "Carbon in Spain: A Black Future." The study showed that the Spanish government imports 24 million metric tons of coal and pays €2,500 million in state aid to the coal industry every year, more than the combined wind, solar, and biomass energy industries receive.[87] One Greenpeace organizer summarized the report this way:

> We published a report on coal where we analyzed the amount of coal produced in Spain, how it is used, and the amount of subsidies being directed to coal. It's interesting to find that coal provides 33 percent of the whole electricity demand in Spain, but up to 64 percent of CO_2 emissions, so it's quite inefficient. And at the same time, we showed that the amount of subsidies for coal in the last eight years were bigger than the support given to all renewables. It is remarkable to know that coal is providing some five thousand jobs in Spain and renewable energy is providing close to two hundred thousand jobs.[88]

Second, Greenpeace together with other environmental groups lobbied for the rapid adoption and strict implementation of FITs. Greenpeace was not only the first NGO in Spain to lobby for the adoption of tariffs, it was also the first to participate as an external consultant in the subsequent revisions, defending the tariffs when they were under threat. As one organizer stated, "We were the main and first proponents of feed-in tariffs for electricity, so we were very closely involved. The feed-in tariff was not passed as a law but as a decree, there have been many reviews and we have been closely involved with those as well. We have been following the issue very closely and challenged any attempts to lower the feed-in tariffs. We are now developing proposals to make the system a law so that there can be more regulation."[89] He goes on to explain why Greenpeace supports introducing legislation on FITs: "It's good to remember that the current regulations came after a strong fight between some companies and organizations that are in favor of wind power and others that are not. This happened every time there was a new regulation in the system. We believe that the best way to avoid fights is to set up a law that sets the feed-in tariff definitely, and establishes a clear framework not just for wind power but all renewables."[90]

Third, realizing that any significant reduction in greenhouse gases depends on the public's acceptance of new technologies such as wind power, environmental activists engaged in significant public-education efforts. As more and more large-scale wind farms were being built around the country, the prospect of local opposition was growing. Spain was particularly exposed to this problem because, unlike in Germany or Denmark, very few wind cooperatives existed in Spain and most wind farms were owned by large companies. Environmental organizations such as Greenpeace, however, were very active in providing public education and publishing reports in support of wind power. For example, Greenpeace was involved in an educational campaign to increase acceptance of wind turbines in 1995 when one of the first wind farms was built near Pamplona, in Navarra. As one government official recognized, Greenpeace's involvement was essential for the local acceptance of the wind farm: "Greenpeace was invited to a week of conferences and seminars in order to discuss the first wind farm. They supported entirely this policy and organized a few public events. For the government and wind companies, it was fantastic to have that support; I think that this collaboration gave a lot of credibility to wind-farm development in Navarra."[91] Not surprisingly, opinion polls show that acceptance of wind energy growth has been getting stronger at the same time as more wind farms have been built.[92]

It is important to note, however, that the environmental movement has been less influential in Spain than in Germany or Denmark, for two reasons. One of them is Spain's dependency on foreign oil, coal, and natural gas. Because of this, the Spanish government created the Institute for Energy Diversification and Saving (IDAE), which adopted a national renewable energy plan focused on reducing the cost of electricity by increasing energy independence. Another reason is Spain's high rate of unemployment, which reached an average of approximately 20 percent in the mid-1990s but was

significantly higher in poor rural areas. Consequently, some regional governments in Spain adopted and implemented pro–wind energy policies first
and foremost in order to address economic problems, and only secondly in
order to address environmental problems. As the head of the European Wind
Energy Association has argued, the recent growth of the wind energy industry
in Spain is, to a large degree, "a story about regional growth, economic
deployment, driving an economy that requires increasing amounts of energy.
There's more of a fundamental value of wind power to an economy in Spain
than in Northern Europe."[93]

The Environmental Movement and Renewable
Energy Policies in Other European Countries

While in Germany, Denmark, and Spain environmental movements had a
strong impact on the wind energy industry, in other European countries
environmental movements' achievements have been less impressive.
Consider the cases of France, Sweden, Austria, and Norway, which have
very good wind potential but smaller wind energy industries than their
above-mentioned neighbors. More specifically, France adopted a renewable
energy FIT policy in 2001 and produced about 0.2 percent of its electricity
from wind power in 2005; Sweden adopted a FIT policy in 1998 and produced about 0.6 percent of its electricity from wind power in 2005; Austria
adopted a FIT policy in 2002 and produced about 2 percent of its electricity
from wind power in 2005; and Norway adopted a FIT policy in 1999 and
produced approximately 0.5 percent from wind in 2005.[94]

France has a relatively weak environmental movement, particularly in
comparison with the German or Danish movements. Although France was
the battleground for some of the earliest antinuclear protests during the
1970s, the issue gradually disappeared from the national political agenda.
Nuclear power expanded rapidly in France because the antinuclear campaign
was poorly coordinated and the political opportunity structure was unfavorable to antinuclear protesters. The French state was strongly engaged in
nuclear energy: it used repressive strategies and created a nuclear
administration with a strong implementation capacity. Furthermore, the
French mass media frequently distorted facts and events and was biased
against antinuclear demonstrators (Rucht 1994). As a result, by the end of
the 1970s the French antinuclear campaign was in disarray and the issue of
nuclear power had lost prominence.

The French environmental movement remained weak during the 1980s
and 1990s. Not even the Chernobyl disaster in 1986 was able to galvanize
the environmental movement and generate mass protests or change public
opinion on nuclear energy (Rucht 1994). The environmental movement also
failed to mobilize the French population on non-nuclear issues. One study
found that the number of environmental protests reported in major French

media sources between 1988 and 1997 ranged between twenty and forty per year (Fillieule 2003). In contrast, in Germany the number of environmental protests reported in major German media during the same period ranged between fifty and two hundred per year (Rucht and Roose 2003). Furthermore, French environmental groups have rarely challenged the state through direct action such as demonstrations and protests because the groups are weak and financially dependent on subsidies from officials and public administrations. As one author notes, it is particularly striking that demonstrations on environmental issues are rare in France because demonstrations are exceptionally common in this country (Fillieule 2003, 75).

The weakness of domestic environmental organizations, combined with a lack of influential political allies and an unfavorable mass media, contributed to the fact that the French environmental movement could neither prevent the government from implementing its ambitious nuclear energy plan nor develop an alternative energy plan during the twentieth century. Not surprisingly, at the beginning of the twenty-first century, France produced almost 80 percent of its energy from nuclear power—the highest percentage of any country in the world.[95]

The French environmental movement, however, had some influence on the French government's decision to adopt a FIT through a campaign against global climate change organized by domestic and transnational environmental organizations. The growing pressure put by environmental activists on EU states to address climate change resulted in the adoption of the Directive on Electricity Production from Renewable Energy Sources in 2001, the first EU directive for promoting renewable energy use in electricity generation. As one study noted, transnational environmental groups such as Greenpeace and the World Wildlife Fund contributed to the adoption of the Directive on Electricity Production from Renewable Energy Sources because they "very resolutely espoused the cause of renewable power, conducting important campaigns at a time when the renewable power industry itself was still in an early phase of its development" (Lauber 2005, 206). The adoption of the FIT by the French government in 2001 was partly a response to this directive, which required France to increase its share of renewable energy from 15 percent in 1997 to 21 percent in 2010. After the adoption of the tariff, the wind energy industry expanded rapidly, and by early 2009 France had a wind power installed capacity of over 3.4 GW.

The Swedish environmental movement also had a more modest impact on the wind energy industry than the German or Danish movements. The antinuclear campaign in Sweden gained popularity when a variety of large and small groups formed the Action Stop Nuclear Energy (ASK) organization in Stockholm in 1974. With support from the Social Democrat Party, antinuclear campaigners established adult-education associations that disseminated the arguments of antinuclear campaign intellectuals to the public (Flam and Jamison 1994, 176). By 1975, however, the antinuclear campaign had split apart between supporters of Björn Gillberg—a prominent environmental leader sometimes described as Sweden's Ralph Nader—and his rivals (Jamison, Eyerman, and Cramer 1990).

The split within the Swedish movement, and the organizers' tactical decision to form an "inter-political popular front" that limited argumentation to the common technical-economic denominator, undermined the campaign's ability to challenge the government's ambitious nuclear energy program (Flam and Jamison 1994). In 1978 the People's Campaign against Nuclear Energy was formed to focus the antinuclear campaign on a national referendum. However, this new organization lost the 1980 referendum on the future of nuclear energy in Sweden not only because of its tactical mistakes but also because of the government's strategy of organizing the referendum on three "lines" organized around established politics rather than on two clearly defined alternatives—a decision perceived by many as a strategic sham (Flam and Jamison 1994). Consequently, the impact of the Swedish antinuclear campaign on Sweden's energy policy was mixed. On the decision side, the campaign did contribute to a minor reduction in the number of nuclear power plants, when the Swedish parliament reduced the number of planned reactors from thirteen to twelve in 1979. On the implementation side, the campaign did not have a significant impact; as one study notes, "once the policy course was set on twelve reactors for Sweden in 1979, no deviations from this course could be observed on the implementation side" (Flam and Jamison 1994, 194).

The antinuclear campaign's limited success in inhibiting the growth of the nuclear power industry partly accounts for the fact that Sweden did not make renewable energy a national priority for a long time and that it adopted a renewable energy FIT only in 1998. Another factor that contributed to the relatively late adoption of this policy is the fact that Sweden has abundant hydropower resources. Indeed, at the beginning of the twenty-first century, Sweden produced almost 46 percent of its electricity from hydropower and another 46 percent from nuclear power.[96] Therefore, Sweden has a relatively small carbon footprint per capita, and it did not witness massive campaigns against acid rain or global climate change.

Yet another factor that contributed to the relatively late adoption of the FIT in Sweden is that, while renewable energy was researched and promoted by a handful of engineers and architects in rural communes and a few of the people's high schools, "such activities have never become particularly significant within the Swedish context" (Jamison, Eyerman, and Cramer 1990, 48). Therefore, no major Swedish manufacturers of wind turbines emerged to lobby for the adoption of pro–renewable energy policies before the beginning of the twenty-first century. Nevertheless, because Sweden has integrated environmental concern into its social and cultural life, and because it has assumed a leadership role in international environmental politics (Jamison 2001), the Swedish government has vowed to address global climate change not only by pressuring other states to reduce their emissions of greenhouse gases but also by promoting renewable energy at home. In fact, in 2006 the Swedish government announced that it will promote renewable energy such as wind and biofuels and try to "wean itself off oil completely within 15 years—without building a new generation of nuclear power stations."[97]

While space restrictions do not permit a detailed analysis of the environmental movement's role in the adoption of FITs in all European countries, it is worth mentioning that the movement also had a moderate influence on the adoption of these policies in other countries. In Austria, a vigorous antinuclear campaign had a significant influence on the government's decision to abandon the construction of nuclear power plants (Preglau 1994). Yet, because Austria has abundant water power, the government focused mostly on promoting large-scale hydropower plants rather than developing wind energy. The situation changed in 2002 when, under pressure from domestic activism against global climate change and from the EU Directive on Electricity Production from Renewable Energy Sources, the Austrian government decided to support wind and solar power through a FIT. Similarly, in Norway sustained mobilizations against nuclear energy and tremendous water-power resources contributed to the fact that the Norwegian government rejected nuclear power and promoted mainly hydropower (Andersen and Midttun 1994). Thus, Norway produced almost 98 percent of its electricity from hydropower at the end of the twentieth century and emitted very few greenhouse gases per capita. Because its government assumed a leadership role in international environmental politics, the Norwegian government also decided to promote renewables such as wind power by adopting a FIT in 1999.

Conclusion

This chapter has examined the ways in which environmental groups and activists shape the energy policymaking processes. Based on case studies of countries that adopted early and strong renewable energy FIT policies—Germany, Denmark, and Spain—it has shown that the environmental movement contributed to both the adoption and implementation of those policies through campaigns against nuclear power, air pollution, and global climate change. The chapter has also shown that environmental groups' ability to influence the adoption and implementation of pro–renewable energy policies depends on their ability to mobilize large green-energy advocacy coalitions, to take advantage of favorable political contexts and a committed and unbiased mass media, and to instill positive public opinion.

Starting in the 1970s, European environmental groups began to mobilize against government plans to build nuclear power plants in response to energy crises. When antinuclear activists organized sustained, large-scale campaigns, and when they had influential political allies, their actions played an important role in the adoption of moratoriums on new nuclear power plants. The antinu-clear campaigns also resulted in the emergence of grassroots groups and research institutes dedicated to promoting the use of clean, renewable energy. These organizations developed key ideas that formed the foundation for renewable energy FIT policies in countries such as Germany and Denmark. In other coun-tries—for example, France or Sweden—antinuclear mobilizations did not result in the emergence of renewable energy grassroots organizations and research institutes. Consequently, these countries developed massive nuclear energy industries and did not view renewable energy as a viable alternative until the end of the twentieth or the beginning of the twenty-first century, when they finally adopted FITs.

During the late 1980s, global climate change became the dominant issue on environmental groups' agendas. Arguing that massive investments in wind power and other forms of renewable energy are essential for address-ing this problem, environmental groups in Germany, Denmark, and Spain relentlessly advocated for the adoption of pro–renewable energy policies. They also vigorously defended their implementation whenever the policies were threatened by the fossil-fuel, nuclear energy, or utility lobbies. Environmental groups fighting for strong renewable energy FIT policies have been successful, particularly when they could build large pro–renewable energy coalitions with unions, farmers, and civic associations; when they have had allies among political elites; and when mass media coverage of environmental issues has been favorable and public opinion positive. The next chapter will examine the adoption and implementation of renewable portfolio standards and other policies in countries where the wind energy industry has developed more slowly.

3

Environmental Campaigns and the Adoption and Implementation of Renewable Portfolio Standards

All I say is "Give us a small fraction of the subsidy that the fossil fuel and nuclear industries are getting and we'll solve the electricity and environmental problems of the world!" The technology is working, the prices are coming down; we are going to have more wind and solar. The question is, in the next five years, do we throw the dollars down the toilet in outdated coal plants, or do we right now act like we can see the future? So far, politicians voted badly. We will probably waste several more billion dollars; they will blame environmentalists for the cost of dealing with global warming. The truth is that wind power is barely more expensive now than coal power. Why build plants that you are going to have to retrofit or retire in a few years?
—Environmental Defense organizer, December 2007

Modest Results: Renewable Portfolio Standards and other Policies in the United Kingdom, United States, and Canada

The United States started developing a wind energy industry relatively early but failed to achieve sustained growth. In 1990 the total installed nominal capacity from wind power in the United States was 1.485 GW, significantly higher than in all European countries combined. By 1997, however, the United States had installed only 1.611 GW and was falling behind Germany.[1] In 2007, despite a record growth of 5.244 GW, the total installed nominal capacity from wind power was 16.818 GW, representing almost 1 percent of the national electricity supply.[2] In the same year, in the United Kingdom wind energy supplied slightly more than 1 percent, while in Canada it supplied slightly less than 1 percent of the electricity demand.

Why did the United States lose its position as a world leader in wind energy during the 1990s? Why do the United Kingdom and Canada lag behind Germany and Spain, despite having some of the best wind resources in the world? This chapter examines efforts to adopt and implement a renewable portfolio standard (RPS) and other policies in the United Kingdom, United States, and Canada, countries that have very good wind potential but somewhat underdeveloped wind energy industries. Rather than arguing that these countries adopted RPS policies—instead of renewable energy feed-in tariffs (FITs)—because of their neoliberal ideological traditions, the chapter shows that the adoption and implementation of RPS policies was a contentious process shaped by environmental groups and activists through the mobilization of large renewable energy advocacy coalitions.[3] These mobilizations had a lower impact on the national and regional energy policies in these countries than mobilizations had in some of the countries examined in the previous chapter because they were less sustained and encountered less-favorable social contexts. Recently, however, reinvigorated climate change campaigns and more-favorable social and political contexts have affected the implementation of RPS policies and contributed to the introduction of a national renewable energy FIT policy in the United Kingdom; they have also resulted in local FIT policies in the United States and Canada.

An RPS, also named a renewable set-aside, mandates that a certain percentage of all electricity be generated from renewable sources. These policies acknowledge that renewable energy technology is not mature enough to be able to compete on the open market, and they create a separate market within which renewable energy projects compete among themselves. By encouraging competition among all renewable technologies in a reserved market, RPS policies aim to reduce costs and stimulate technological development. However, because renewable energy technologies are in very different stages of development, open competition among these technologies often results in market domination by a few technologies (Redlinger, Andersen, and Morthorst 2002, 175). To avoid this, some RPS policies further allocate specific percentages of the renewable market for specific technologies. Usually, RPS purchase obligations increase over time, and retail suppliers must demonstrate compliance on an annual basis. Many RPS policies are mandatory—they are backed by various types of compliance enforcement mechanisms—and include the trading of renewable energy certificates.

The Environmental Movement and Renewable Energy Policies in the United Kingdom

Much of the growth of the wind energy industry in the United Kingdom before 2008 can be attributed to RPS policies. In 1990, the United Kingdom adopted the Non–Fossil Fuel Obligation (NFFO). This policy was primarily intended as a support scheme for nuclear power plants, which could not survive in the newly privatized, competitive electricity market. The NFFO also aimed to reduce the price of renewable electricity generation by creating separate competitive markets for specific renewable technologies. However, due to a number of problems with planning regulations and overestimates in the price reductions of renewable energy projects, the NFFO was not very successful in stimulating the development of wind and other renewable energy technologies.[4] A new policy, called the Renewables Obligation (RO), was adopted in 2002 to replace the NFFO. Similarly to the RPS policies adopted by some U.S. states, the RO mandated that electricity generators supply a percentage of their electricity from renewable energy. More specifically, the RO required licensed electricity suppliers to source 9.1 percent of electricity from renewable sources in 2008/2009, and 15.4 percent in 2015/2016.

In 2008 the United Kingdom became the latest country in the world to adopt a FIT. Compared to the German or Spanish tariffs, which have no project size limits, the British policy was modest: it applied only to projects up to 5 MW. It also lacked the specific provisions or prices that are part of the German tariff; instead, specific provisions were determined administratively in 2009. Nevertheless, renewable energy advocates consider it an impressive achievement not only because it represents "an ideological breakthrough"—a departure from the British government's previous support for quota systems as the only mechanism for developing renewable energy— but also because it is expected to have a major future impact on renewable energy industries in Britain.[5] According to the climate change minister, for example, "This decision means that installing equipment like wind turbines, solar panels or biomass heaters will be much more financially attractive. It will make a real difference to families, communities and businesses that want to generate their own energy."[6]

What was the role of the British environmental movement for the adoption of the RPS and FIT policies? In contrast to the antinuclear campaigns in other European countries, the antinuclear campaign in the United Kingdom has been relatively inconsequential. Although the British environmental movement expanded considerably during the 1970s, it did not mobilize effectively on the issue of nuclear power. Local opposition to nuclear power plants during the 1970s was weak and did not include large coalitions between environmental groups, farmers, and grassroots associations. Public protests were relatively small—the biggest antinuclear demonstration attracted only twelve thousand people—and did not impact the national elections. Moreover, the Green Party has been the only national party firmly committed to closing nuclear power plants, but its impact on this issue has been negligible (Rüdig 1994, 91).

Throughout the 1980s and 1990s, British environmental organizations continued to grow. Membership in the largest and oldest environmental organizations doubled between 1981 and 1991; at the same time, the number of members or supporting donors of Friends of the Earth (FOE) grew sixfold and those of Greenpeace grew tenfold. By 2000, the Royal Society for the Protection of Birds alone had more than one million members (Rootes 2003, 20, 22). Yet the antinuclear issue was not at the top of the agenda for most environmental groups. One study found that this issue ranked only fifth among the main issues of protest for U.K. environmental groups between 1988 and 1997, after transportation, animal welfare and hunting, nature conservation, and urban and industrial pollution. Expressed as a percentage of the total number of environmental protests in the United Kingdom during this period, nuclear energy and nuclear waste accounted for less than 8 percent; in contrast, in Germany they accounted for over 45 percent, while in Spain they accounted for almost 17 percent.[7] In fact, the environmental movement has had less influence in the United Kingdom than in other European countries not only on the issue of nuclear power but also on recycling. For example, in 2002 the United Kingdom recycled about 12 percent of its waste, while most European countries recycled well over 33 percent of their waste.[8]

The weak presence of the nuclear power issue on environmental groups' agenda helps explain why the United Kingdom was using nuclear power to produce a large proportion (approximately 25 percent) of its electricity at the beginning of the twenty-first century. Additionally, the United Kingdom's antinuclear campaigners did not have influential political allies who supported nuclear power moratoriums.[9] But two other factors are equally or even more important for nuclear power's significant contribution to the United Kingdom's energy supply. First, much of the nuclear capacity was installed in the 1960s, before the formation of the modern environmental movement. Second, the United Kingdom's need for additional nuclear capacity was greatly reduced when cheap and plentiful natural gas was discovered in the North Sea in the early 1990s.[10] Consequently, the antinuclear campaigns in the United Kingdom did not result in the search for alternative sources of energy on the same scale as in Germany, Denmark, and Spain.

Environmental organizations, however, have been able to exert some influence on the energy policymaking process through their recent campaigns against global climate change. Greenpeace, FOE, and other British environmental groups campaigned against global climate change starting in the early 1990s. Most environmental groups used tactics such as protests, petitions, and boycotts in order to build governmental support for vigorous climate change action and to pressure businesses to reduce their carbon footprint. A grassroots network of environmental activists committed to direct action was formed in 2000 under the name Rising Tide; its call to action demanded "an immediate end to oil exploration and a dismantling of the fossil fuel economy."[11] In 2001, Greenpeace and FOE launched a major campaign to try to force Esso, and its American parent Exxon, to abandon their opposition to the Kyoto Protocol. In 2005, numerous British environmental groups formed the Stop Climate Chaos Coalition, an umbrella organization that currently includes over one hundred organizations, "from environment and development charities to unions, faith, community, and women's groups."[12] In November 2006, the coalition organized the iCount London Climate Change Rally, a demonstration attended by almost twenty-five thousand people and timed to coincide with the release of the Stern Report, which called on the government to take more serious action to prevent damage from climate change.[13]

One of the organizations that has had a major influence on the climate change debate in the United Kingdom is FOE. The organization launched the Big Ask campaign in 2005, which asked the government to commit to a significant reduction in the amount of carbon dioxide being released every year, or a cut of 80 percent by 2050. To build support for a parliamentary petition calling for a new law requiring annual cuts in carbon dioxide emissions of 3 percent, FOE recruited Thom Yorke, the front man from the well-known rock band Radiohead. In 2006, Radiohead launched the Big Ask Live concert; FOE distributed information through the Internet, in cinemas, and in print media, and collected signatures from all over the country. A key element of FOE's strategy was that it did not stop at building public support for climate action by organizing festivals, public meetings, and concerts; it also directly engaged policymakers. The organization drafted the Climate Change Bill and got three MPs to sponsor it, then asked MPs to support an "early day motion" to get the bill into the statute book. In 2007, the government adopted the Climate Change Bill as part of its legislative program, but FOE pressured the government to strengthen the bill by asking its members and sympathizers to contact their MP.

The Big Ask campaign was a long but ultimately successful effort. According to FOE, over two hundred thousand people contacted their MP, and 620 out of 646 MPs were asked by constituents to support the campaign for a strong climate change law. Toward the end of 2007, FOE increased its pressure and launched the Big Ask online march so people could lobby their MP by video message. At the beginning of 2008, the U.K. government announced that the Climate Change Bill would set annual milestones for emissions reductions, and by the end of the year, MPs had voted in favor of the Climate Change Act 2008, which included all U.K. emissions and would cut greenhouse gases by 80 percent by 2050.[14] Friends of the Earth claims that this is a direct outcome of its campaign: "This is a massive success for Friends of the Earth's Big Ask campaign. We couldn't have done it without the thousands and thousands of people who contacted their MP to ask for just such a bill."[15] The secretary of state for energy and climate change also recognized the crucial role played by the environmental movement and by the members of the public who contacted their MP: "I pay tribute to the scientists who detected the problem, the campaigners who fought to bring it to public attention, the green movement that mobilized for change, and above all, the members of the public who wrote to us in record numbers, asking for a bill that met the scale of the challenge."[16]

The major success of the FOE campaign, however, depended also on the presence of numerous political allies. Many Conservatives and Liberal Democrats supported a measure to set, monitor, and enforce annual cuts in greenhouse gas emissions. As David Cameron, the Tory leader, said, "Climate change is one of the greatest challenges facing us today, and we can only tackle it if we realize that we all have a responsibility to act—individuals, businesses and government. The Government must deliver a proper climate change bill in the Queen's speech—not a watered-down version." And the environment spokesman for the Liberal Democrats stated that "the climate change bill has to have teeth and bite. There must be annual targets that allow the Government's progress on climate change to be assessed." In fact, a total of 412 MPs, including 202 Labour MPs, signed a House of Commons motion calling for a 3 percent cut in emissions each year. As Tony Juniper, the director of FOE, acknowledged, "There is now overwhelming cross-party support for new legislation to cut U.K. carbon dioxide emissions by at least 3 percent every year. We hope that ministers will seize the opportunity presented by this political consensus and make the U.K. a world leader in developing a low carbon economy."[17]

Greenpeace is another environmental organization that has campaigned hard in the United Kingdom on the issue of climate change. Greenpeace activists have repeatedly staged anticoal protests that have attracted the mass media's attention. In 2007, Greenpeace activists organized perhaps one of the most consequential climate change protest events when they managed to shut down a coal-fired power plant in Kingsnorth and, after being arrested and charged with criminal damage, were acquitted based on the argument that they prevented future property damage from climate change. The trial was covered extensively in the British media partly because of the defense's strategy. The defense called as a witness James Hansen, one of the world's leading climate scientists. Hansen told the court that more than a million species would be made extinct because of climate change and calculated that the Kingsnorth plant would proportionally be responsible for four hundred of these extinctions. He stated he agreed with Al Gore's statement that more people should be chaining themselves to coal-powered stations, and declared, "Somebody needs to step forward and say there has to be a moratorium, draw a line in the sand and say no more coal-fired power stations."[18]

Greenpeace and FOE have committed not only to "high-profile critical campaigning" but also "to carrying out or funding research on environmental issues and, increasingly, to 'solutions campaigning' designed to promote better environmental practice" (Rootes 2003, 22). In the late 1990s, FOE conducted a study to assess the success of the NFFO policy in Scotland. The study found that the NFFO was failing to deliver because of a low take-up rate—for example, of all the projects that had been awarded government contracts since 1994, only 21 percent were generating electricity by 1999; moreover, about 25 percent of the first round of contracts was still stuck in the planning stage.[19] Realizing that the NFFO policy had failed to sufficiently stimulate the growth of the wind and other renewable energy industries, FOE and other environmental organizations called for a new policy: the RO. They also mobilized to remove one of the main barriers to wind project development: local opposition by not-in-my-backyard (NIMBY) groups such as Country Guardians. Thus, FOE, Greenpeace, and the World Wildlife Fund launched Yes 2 Wind, a website that provides information and resources for increasing public support for wind farms. In 2005, local environmental activists with ties to these organizations also formed the Sustainable Energy Alliance, a group that aims to disseminate information about and build local support for new wind projects in Wales.

In addition to the impact they had on the government's decision to adopt the Climate Change Bill and the RO policy, environmental groups campaigning for climate change action have attempted—with mixed results—to influence the British government's annual Energy Policy Review. The 2002 review by the Prime Minister's Strategy Unit was notable for its dismissal of nuclear power and support for renewable energy. However, because the nuclear power lobby mobilized vigorously and pushed for reconsideration of this review, the subsequent energy review published in 2006 stated that "new nuclear power stations would make a significant contribution to meeting our energy policy goals."[20]

The environmental groups fought back, and at the end of 2006, Greenpeace initiated a judicial review process. In court, Greenpeace complained that there had been a failure to present clear proposals and information on key issues, such as disposal of radioactive waste and building costs. The review process ruled that elements of the 2006 Energy Policy Review were "seriously flawed" and "not merely inadequate but also misleading."[21] The government responded by announcing at the beginning of 2007 a public consultation on whether to build new nuclear power stations; yet Greenpeace, FOE, and four other environmental organizations withdrew from the consultation in September 2007, accusing the government of a "public relations stitch-up."[22] In January 2008, the U.K. government gave the go-ahead for a new generation of nuclear power stations to be built, and despite the increasing ferocity of the fight against new nuclear power plants, some environmental activists were prepared to endorse nuclear power in order to reduce greenhouse gas emissions.[23]

Environmental groups involved in climate change campaigns have contributed to the adoption and implementation not only of the RO policy but also of the FIT. In early 2008, environmental activists from FOE, Greenpeace, and the Renewable Energy Association (REA) launched a FIT campaign. The NGOs started the campaign by advertising in three national newspapers (the *Independent*, the *Times*, and the *Guardian*) the need to adopt a policy that would dramatically increase the use of renewable energy in the country. The advertisements used a European football analogy "to depict the national shame of the U.K.'s feeble performance in renewable energy compared with that of Germany," noting that Germany had two hundred times more solar power and more than ten times more wind power installed than the United Kingdom.[24] As one REA spokesperson put it, "The U.K. certainly has a renewable energy goal to achieve in Europe but we are definitely not scoring. We need exactly this kind of measure [FIT] to move from our frankly embarrassing bottom league position in the EU."[25]

The campaign's main goal was to effect a breakthrough in the United Kingdom's renewable energy industries; in the words of one organizer, "It would mean a green energy revolution literally on the doorsteps of communities all over the U.K. and the starting point for an 'Energy Generating Democracy' where people are not simply consumers of energy but generators of their own heat and power."[26] To make this happen, FOE and its allies lobbied for the adoption of a FIT that had a maximum project size of at least 10 MW—the equivalent of about five modern wind turbines—because larger projects are more attractive to investors. The environmental groups' strategy was two-pronged: on the one hand, environmental activists built public support for the measure by collecting signatures from celebrities and other citizens; on the other hand, they lobbied MPs and built bipartisan support for the measure.

A key element in the success of the FIT campaign was the organizers' ability to build a coalition that included not only all major environmental groups and renewable energy associations but also various associations that represented potential investors in renewable energy. This point is underscored by an organizer from the REA:

> We partnered up with Friends of the Earth because their campaign is most engaged; the guys in the economics team are very, very keen on the tariff. We started working together, and also trying to expand the coalition to include other potential investors. What we sought to do was identify those potential investors and get them campaigning with us. In the end we had groups like the Country Men of Business Association, which represents about half of the land owned in the U.K., we had the Home Builders Federation, which represents 80 percent of the construction industry, we had the National Farmers Union, which represents the majority of farmers, and we had professionals from institutions such as civil engineers and mechanical engineers. We had all the environmental NGOs. We had a commerce association which represents the majority of big retail in the U.K. And so, we involved actors who could invest in renewable energy and who weren't really being incentivized to do that at the moment. It was very effective to go in with interests like that, like a major bank that was willing to invest hundreds of millions of pounds, and say, "Give us the scheme that works for us."[27]

Another important element of the success of the FIT campaign was the organizers' ability to find political allies. For example, one REA member had previous experience in coalition building from working with Greenpeace on renewable energy generation. As she explains,

I did a lot of work before with Greenpeace on decentralized energy and getting the Conservatives on board on that. So that was very helpful, but we also had help from some backbench Labour MPs: in particular a guy called Alan Simpson. He's very green, his own house is fitted with renewable energy and he's been a long-time campaigner and very supportive of the idea of a tariff. We got this on the agenda early on, and the energy minister was put under pressure, and there were some very good debates.[28]

Yet another key element of the campaign was the support from celebrities. According to the same REA campaign organizer:

The government started to be aware that there was very strong cross-party support for this from some very credible people, and that the coalition outside was growing. And I hate to say it as well, but actually celebrity support made a hideous difference. We have a pop singer here; her name is Lily Allen...She supported the campaign, gave it some nice credit and an email went out to all the MPs. Part of this coalition was this recording studio called The Premises which obviously has a lot of links to celebrities. That created a lot of excitement, you know, MPs who were all craving Lily Allen songs, and things like that. It did really help, having a bit more sex appeal.[29]

In the intense political battle that ensued, environmental activists were able to secure cross-party support in Parliament: a FIT amendment was launched by Conservative peer Baroness Wilcox, Labour peer Lord Puttnam, and Liberal Democrat peer Lord Redesdale.[30] The campaign ran into opposition from most electricity producers and, surprisingly, even some members of the renewable energy industry, which feared that competition from small-scale projects would reduce their profits from large-scale projects. Because of a peculiarity of the British RO scheme, a shortfall of renewable energy increased the value of renewable energy certificates and the price paid for supplying renewable energy. Some energy analysts even argue that "the large bulk of the wind power that has been deployed in the U.K. is owned by one or the other of the six electricity supply companies that dominate the U.K. electricity system and who also have an interest in keeping ROC [Renewables Obligation Certificate] prices up by keeping fulfillment down."[31] Alan Simpson, the Labour MP who led the debate in the House of Commons, states,

Many of the big energy suppliers have been fighting tooth and claw to prevent us from doing anything as bold and imaginative as we are doing. The Association of Electricity Producers had lobbied for a threshold of 50 kW. The British Wind Energy Association lobbied, until the last moment, for a threshold of 500 kW. Such demands would preclude the opportunity to develop genuine, transformational renewable energy systems on a community, town or city scale. The Secretary of State should be praised for his determination and willingness to push the boat out much further than many of those vested interests would have felt comfortable with.[32]

The final victory, however, came not only because of the well-organized FIT campaign but also because the political context changed rather unexpectedly. One REA member accounts for the campaign's success, despite significant opposition, this way:

> The coalition we had put together was pretty hard to resist, and then over the summer we just chipped and chipped away at them. The campaign was growing, we had the party conferences, we had fringe events where everyone was talking about the tariff, and then when it started picking up again in the [House of] Lords for the final processes. Then, we were lucky; we had a big stroke of luck with the new Department of Energy and Climate Change coming in. Gordon Brown announced he was going to start this new department for energy and climate change, which the whole of the environmental movement was thrilled about because we'd been calling for that for a very, very long time and it's so we can bring energy efficiency, renewables, everything together under one place, with a very clear link to climate change. And he put in charge a politician who was young, fresh, quite green. Again, because of the politics of this, because Labour's very keen to get a lead on the Tories, who have been doing very well here in opposition, I think they were looking to be more radical. That was a stroke of luck, I think, that he came in at the right time, he looked at this, he got the vision, and he had the courage actually, to pick it up very quickly and run with it.[33]

As a final point, it is important to emphasize that the British environmental groups' ability to shape the adoption of renewable energy policies, as well as the broader debate on energy, was influenced by the availability of political allies and by a public opinion that supported strong action against climate change. The adoption of the RO and the FIT resulted from well-organized campaigns, but greatly benefited from the support it received from some top-level politicians and a large segment of the population. Additionally, the British media covered the climate change issue in positive terms, while the fossil-fuel and nuclear energy lobbies' opposition to renewable energy policies was rather moderate.[34]

The Environmental Movement and Renewable Energy Policies in the United States

In the United States, a total of twenty-seven states plus the District of Columbia had adopted RPS policies by 2008. These policies had a major impact on the growth of the wind energy industry. Approximately 8.9 GW of the nonhydro renewable capacity additions in the United States between 1998 and 2007 occurred in states with RPS policies, and the great majority of the renewable capacity additions (approximately 93 percent) came from wind power. These policies will continue to have a major impact in the future; it has been estimated that, assuming that full compliance is achieved, current mandatory state RPS policies will require the addition of roughly 60 GW of new renewable energy capacity by 2025, equivalent to almost 5 percent of projected 2025 electricity generation in the United States.[35] However, even though legislation to establish a national RPS has been considered by the U.S. Congress since 1997 and the Senate has passed RPS proposals on three separate occasions, the United States has no national RPS policy.[36] The federal government has used production tax credits (PTCs) to stimulate investment in new wind energy projects, but because Congress has allowed the federal wind energy production tax to expire periodically, the U.S. wind energy industry has followed "boom and bust" cycles.[37] In addition, by 2010 six states—California, Illinois, Hawaii, Michigan, Minnesota, and Rhode Island—had introduced renewable energy FIT–type legislation.

The United States has some of the best wind resources in the world, yet it has significantly less installed capacity per capita than many European countries. Moreover, most of the installed capacity is concentrated in a few states. For example, in 2000 only California had installed over 1 GW, while in 2006 only California and Texas had installed over 2 GW of wind power capacity. To understand why this is the case, it is important to examine the U.S. environmental movement's opposition to nuclear power and global climate change and its efforts to shape the adoption and implementation of policies such as RPS, PTCs, and renewable energy FITs.

Similarly to many European governments, the U.S. government responded to the 1970s energy crisis by increasing support for nuclear energy. President Nixon launched Project Independence, which called for a massive expansion of nuclear power to reach energy independence by 1980 "in the spirit of Apollo and with the determination of the Manhattan project"; while in 1975 President Ford announced an extremely ambitious plan to bring 200 nuclear plants online by 1985 and 650 more by the end of the century (Joppke 1993, 53). The U.S. Energy Research and Development Administration (ERDA) spent most of its funds on nuclear power. In 1976, for example, approximately 75 percent of its research and development (R & D) funds were spent on nuclear power; and in that year, the fast breeder program—a fast breeder is a fast neutron reactor designed to breed fuel by producing more fissile material than it consumes—alone was estimated to cost over $10 billion (Joppke 1993, 56, 62).

While the U.S. government's ambitious plans to develop the nuclear energy industry in the early 1970s were not met with the same vigorous grassroots mobilizations as in some European countries, they did not proceed unopposed. One of the first groups that mobilized against nuclear power was the National Resources Defense Council (NRDC). Founded in 1970 by a group of law students and attorneys at the forefront of the environmental movement, the NRDC published a report in 1974 that demanded a drastic tightening of plutonium regulation.[38] The early antinuclear position of the NRDC was not aiming to "price reactors out of business, but to protect public health" (Joppke 1993, 59). Less than one year later, however, the NRDC modified its demand and asked that the final decision over the use of plutonium be delayed for several years. Leading the opposition against the fast breeder program, which was supposed to become the core of the coming plutonium economy, the NRDC used two main arguments. One was that a fast breeder reactor was more dangerous than a light-water reactor because it could explode; the other was that economic costs outweighed the expected benefits. Consequently, the NRDC called for a postponement of this program for at least one decade and for development of energy alternatives.

The Union of Concerned Scientists (UCS) also fought against nuclear power starting in the early 1970s. Founded in 1969 by Massachusetts Institute of Technology professors and students, UCS published numerous reports that were critical of safety problems and functioned as a repository for whistleblowers and nuclear dissenters. The UCS was also at the forefront of the political battle over nuclear power. In 1975, a group of nuclear scientists that included twelve Nobel Prize laureates announced that there was "no alternative to nuclear power" and warned that "the end of our civilization was near unless the United States committed itself to nuclear power to combat the energy crisis" (Joppke 1993, 64). The UCS promptly launched a signature drive endorsed by 2,300 scientists, including nine Nobel laureates, that demanded a drastic reduction in the construction of new reactors. This became a major event because "it destroyed the industry argument that no reputable scientists had doubts about the safety of the reactors" (Joppke 1993, 64).

By the mid-1970s, the U.S. antinuclear movement had grown from a "loose network of concerned individuals and groups spotted throughout the country" to a popular movement (Joppke 1993, 65). The main event that led to this transformation was the introduction of Proposition 15 in California by antinuclear activists. This proposition aimed to gradually shut down existing nuclear plants and prevent the construction of new ones in California unless the industry accepted unlimited nuclear liability and demonstrated that nuclear waste could be stored safely. Proposition 15 and other similar initiatives introduced in six states were overwhelmingly defeated because, as one antinuclear activist put it, "We never found a way to dispel the myths about cheap [nuclear] energy and jobs" (Joppke 1993, 67). Antinuclear activists nevertheless succeeded in defeating two major nuclear projects and in introducing a nuclear moratorium on new nuclear power plants in California.

Antinuclear opposition continued to grow during the late 1970s and early 1980s. In terms of mass mobilization, the antinuclear campaign reached a climax when groups as diverse as the UCS, the Clamshell Alliance, and Critical Mass organized a protest of over seventy thousand people following the Three Mile Island nuclear accident in 1979. Antinuclear activists became more and more involved in politics. For example, FOE, Critical Mass, and the UCS organized the Safe Energy '80 campaign, which asked environmentalists to show up at public appearances of presidential candidates and force them "in a gentle but insistent way, to address the nuclear issue" (Joppke 1993, 139). Although antinuclear mobilizations in the United States did not result in a national moratorium on reactor building, as in Germany, they did contribute to such a sharp decrease in public support for nuclear energy that no new nuclear power stations were constructed for three decades.[39] Starting in the late 1970s, however, the antinuclear movement shifted its focus from energy to weapons. In a clear sign of this change, in 1978 the UCS abandoned its primary focus on nuclear energy and started a new campaign to stop the nuclear arms race.[40]

During the 1980s, the U.S. environmental movement shifted the focus of its campaigns from the nuclear energy issue to problems such as acid rain, ozone depletion, and climate change, while also expanding its base. Membership in most large environmental organizations increased dramatically. In the early 1980s, the Sierra Club had 246,000 members, Environmental Defense had 46,000, and the NRDC had 40,000. By 2003, the Sierra Club had 730,000 members, Environmental Defense had 350,000, and the NRDC had 450,000.[41] The increase in membership, combined with the complexity of legislation and constraints from private funders and the government, resulted in increased pressure on environmental groups to formalize and professionalize. Many of these organizations became staffed by scientists, lawyers, and professionals, and participated in activities such as lobbying, litigating, and electioneering. Many of them also hired large staffs. For example, in 1980 the Sierra Club had 145 staff members, Environmental Defense had 41, and the NRDC had 77; yet in 2000, the Sierra Club had 290 staff members, Environmental Defense had 230, and the NRDC had 195 (Bosso 2005, 92).

But perhaps the most important change for the U.S. environmental movement was in tone, not size. As Joppke (1993, 71) notes, in order to broaden its base of support the movement had to "overcome abstract negation and demonstrate the viability of a 'safe energy' alternative." One of the most important advocates of alternative energy was Amory Lovins. Working for the newly founded FOE, Lovins published a series of articles and books that sought to demonstrate the social and economic attractiveness of "soft-energy systems." His idea of a soft-energy path, which relies on energy conservation and renewable energy sources, was "gladly accepted by all movement factions," and it unified various "environmentalist groups under the common banner of safe energy" (Joppke 1993, 72). Unlike the hard-energy path followed by the existing energy sector, which wastes almost two-thirds of spent fuel, the soft-energy path is more efficient because it avoids conversion and distribution losses of energy. Furthermore, Lovins argued, because conservation and renewable technologies are labor- rather than capital-intensive, this path also produces more jobs.[42]

Since the 1980s, a growing number of U.S. environmental organizations and institutes have conducted research and published materials aiming to demonstrate that renewable energy is a practical alternative to nuclear power and fossil fuels. One of those is the Rocky Mountain Institute (RMI). Founded in 1982 by Amory Lovins, the RMI conducts research on energy policy and proposed solutions that encourage energy efficiency and renewable energy. Another is the World Resource Institute (WRI). Also founded in 1982, the WRI publishes regular reports that promote the use of renewable energy. The UCS also promotes renewable energy and wind power in periodic reports. For example, in one of its reports published in 1980, the UCS argued that "the amount of attention and money being devoted to wind technology development by the federal government is astonishingly low given the technological maturity, economic attractiveness, and favorable environmental features of wind generation."[43]

Relatively large and wealthy national environmental organizations such as the UCS, the NRDC, and the WRI contributed not only to the gradual decrease in popular and political support for nuclear energy at the end of the 1970s and in the beginning of the 1980s but also to a modest increase in federal support for renewable energy. Because of a favorable political context during the Carter administration, antinuclear and environmental groups involved in the propagation of a clean energy alternative were able to temporarily influence the federal policy on wind power and other forms of renewable energy. Indeed, the lobbying, litigating, and information-disseminating efforts of these groups contributed to two major changes in the federal energy policy. One was the adoption of the Energy Tax Act of 1978, which provided a 10 percent federal tax credit—raised to 15 percent by the Crude Oil Windfall Profits Act of 1980—on new investment in capital-intensive wind and solar generation technologies.[44] Another one was an increase in R & D funding for renewable energy. The U.S. Department of Energy's funding for renewable energy R & D for the period 1978–1981 was close to the R & D amount spent for fossil fuels, although somewhat smaller than the amount spent for nuclear energy. (In contrast, the Department of Energy's renewable energy R & D for the period 1982–1990 was much smaller than that of fossil fuels and, in particular, nuclear energy.)[45]

The gradual shift away from the nuclear and fossil-fuel technologies and the consequent development of the wind energy industry in some parts of the country was shaped not only by national environmental organizations but also by small local groups. Beginning in the 1970s, environmentalists and renewable energy enthusiasts formed grassroots groups that sought to apply E. F. Schumacher's "Small is beautiful" ideas to energy generation. One such group was the New Alchemy Institute in Massachusetts. The institute was formed in 1971 by a few idealists dedicated to researching technologies that "emphasize a minimal reliance on fossil fuels and operate on a scale accessible to individuals, families and small groups."[46] Members of the institute were wind energy enthusiasts who built 25 kW wind turbines because they wanted to prove that wind power was an alternative to nuclear power. While their wind turbines were small and did not have the nationwide impact of the Tvind wind turbine constructed by the environmental activists in Denmark, their movement had an important role in attracting and training young people such as Tyrone Cashman, who later turned out to be the main architect of California's wind energy policy (Righter 1996).

Environmental groups have also shaped the energy policies of individual states since the late 1970s and early 1980s. The vibrant environmental community in California was instrumental in the great "wind rush" during the early 1980s. Because California provided a hospitable atmosphere to environmentalists and renewable energy enthusiasts, wind energy people from around the country came to "build something together, like the Amish at a barn raising."[47] Tyrone Cashman, who became enchanted with wind power while working at the New Alchemy Institute, was one of those activists. After he joined the Office of Appropriate Technology, Cashman pushed for the introduction of a wind energy investment tax credit in California. The investment tax credit allowed individuals and companies that invested in wind power plants between 1981 and 1985 a 25 percent deduction from their federal income tax, in addition to a federal investment tax credit and business energy investment credit, which offered wind energy companies the opportunity to recoup 25 percent of their investment. The Office of Appropriate Technology also stimulated investment in wind power by publicizing that fossil-fuel costs "make the production of electricity from wind energy one of the most attractive and cost-effective alternative generation technologies currently available" (Righter 1996, 208). The combination of a strong environmental community and a favorable political context (due to the election of a pro-environment governor, Jerry Brown) made California a wind power leader both nationally and internationally during the early 1980s.[48]

Starting at the end of the 1980s, nuclear energy was gradually replaced by global climate change as the most pressing issue for the environmental movement in the United States. Global climate change entered U.S. public and political debate in 1988 when a major drought gripped most of the United States and James Hansen testified before Congress that the earth was already experiencing global warming. Since then, more and more environmental NGOs have begun campaigning against global climate change. Many of them joined the U.S. Climate Action Network, an umbrella organization that focuses primarily on pressuring the U.S. government to ratify an international climate change treaty. In 1997, for example, the UCS circulated a petition titled "A Call to Action," which called for ratification of the Kyoto Protocol. It was signed by 104 Nobel Prize–winning scientists.

Unlike their European counterparts, however, U.S. environmental organizations have had almost no success in shaping the national government's position on climate change. For many years, environmental organizations based in the United States fought a losing battle against the powerful fossil-fuel industry to influence the federal policy on climate change.[49] The U.S. federal government rejected any national policies to address global climate change, while the House of Representatives banned the use of federal funds for activities that could be seen as "back door implementation" of the Kyoto Protocol.[50] Accordingly, in 2008 the United States remained one of a handful of countries that had not ratified the Kyoto Protocol despite well-coordinated environmental campaigns that had lasted for more than two decades.

Although the climate change campaign failed to persuade the U.S. government to sign the Kyoto Protocol, it had some success in shaping two federal energy policies. Realizing that it was the only feasible policy to stimulate the development of renewable energy industries in the United States, many environmental organizations formed a coalition that supported the introduction of production tax credits (PTCs). The idea for PTCs was developed by the American Wind Energy Association at the beginning of the 1990s as a way to address the shortcomings of the investment tax credits and to attract capital for wind and other renewable energy projects. The American Wind Energy Association was joined by "virtually every major environmental group in the country" in its call to the House Tax Committee to include a production-based tax credit to create greater parity in the federal tax code.[51] A PTC targeted to support only wind and certain bioenergy resources was included in the Energy Policy Act of 1992 after considerable lobbying by environmental groups and renewable energy associations on the one side, and fossil-fuel industries and electricity companies on the other side. The credit, however, was relatively modest—it had a value of $0.015/kWh—and it was short term, being set to expire in 1999.

The UCS, the NRDC, the Sierra Club, and other environmental groups have renewed their efforts to extend the PTC a number of times since the early 1990s. In 1999, after intense lobbying efforts, a coalition between environmental organizations, renewable energy associations, and some utilities achieved an extension of the credit until the end of 2001. The coalition mobilized again in 2001 and was able to extend the credit again from March 2002 until December 2003. But in 2003 and 2004, the coalition was defeated; according to the UCS, "from late 2003 through most of 2004, attempts to extend and expand the PTC were held hostage to the fossil-fuel dominated comprehensive energy bill that ultimately failed to pass during the 108th Congress."[52] It was only toward the end of 2004 that a one-year extension of the PTC was included in a larger package of "high priority" tax incentives for businesses after much lobbying by environmental NGOs and their allies. The tax incentives were again set to expire at the end of 2008, but after additional efforts by the pro–renewable energy coalition, they were extended as part of the Emergency Economic Stabilization Act that President Bush signed in October 2008.

The failure to adopt a long-term PTC policy cannot be attributed to a weak or disorganized mobilization effort by environmental groups. Many large U.S. environmental groups used well-coordinated strategies to lobby members of Congress for more than fifteen years to support a PTC. Here is how one Sierra Club organizer describes the organization's efforts to renew renewable energy tax credits:

We spend a lot of our time on Capitol Hill talking to members of Congress. We also have a pretty broad network of grassroots organizers and our members that we've recruited to talk directly to their own elected officials. We've got folks here in D.C., but probably even more important are the over one million supporters that we have across the country who we can get information from about renewable energy legislation and ask to weigh in on their member of Congress. Last year, when the energy bill was being debated, we had a very big grassroots campaign and we generated petition signatures from our supporters, we did "patch-through" phone calls, where we would call up one of our members, tell them about the bill, and ask if they would like to be connected to their member of Congress. We also had meetings across the country, touring renewable energy facilities in different towns, to try and get the word out as much as we could.[53]

The importance of building a coalition that is as inclusive as possible in order to overcome opposition from the fossil-fuel lobby was also emphasized by a member of Environment America. In his words,

Because federal renewable energy production tax credits are not mandates or standards, we have a really broad base of support. We have the utilities, groups like Florida Power and Light, PG&E, but we also have chemical companies, like Dow and DuPont Chemical, manufacturers like Siemens and Owens Corning, and all these groups are interested because this is an incentive to do the right thing and to sell another product. Our coalition is really broad and we've been working very closely within this coalition because it's something that we don't always have, it's new and interesting. We've also been identifying some Republicans who voted for this in the past and kind of put the pressure on them to figure out a way to get this passed, especially some of the Republicans who could be up for reelection this year and could be facing a serious challenge. We hope to get them going to their leadership and say, "We need this to happen, the wind industry is huge in our state and we are going to get hit very hard if the Republicans hold up this bill."[54]

The United States' failure to adopt a long-term PTC policy resulted from an unfavorable political context and from significant opposition from the oil and gas industries. Because the tax credits were supposed to be paid by reducing some of the subsidies that the oil and gas industries were receiving, the legislation was met with fierce opposition from politicians with ties to these industries. According to one Sierra Club organizer, "On the renewable energy tax credit, the legislation would have paid for the tax incentives by rolling back subsidies to the oil industry, so on that particular piece of legislation we've had all of the big oil companies and the oil industry lobby against it really aggressively, and they continue to do so."[55] An organizer from Environment America also emphasizes the role of the oil and gas industry:

On the tax credits, because they have been tied to the oil and tax subsidies, our biggest opponents have been the big oil and the big gas companies. So, if you open up the papers in D.C. you'll see tons of advertisements from all the different oil companies saying "Oil is good for America, so we deserve subsidies like everybody else," things like that. And they've done a pretty good job of labeling the subsidy rollbacks that are in the bill as taxes on their industry that would decrease domestic production of oil, because one of the rollbacks is tied to the incentives for domestic production. Of course, with oil selling for over one hundred dollars per barrel, one would wonder what more subsidies do they need for production?[56]

Although environmental groups were unsuccessful in their attempt to create a coherent national renewable energy policy in the United States, they had some success in introducing an RPS. In fact, one environmental organization—the UCS—initiated the debate about an RPS in the mid-1990s. The first discussions of the detailed design of an RPS began in California. In 1995, the UCS and the American Wind Energy Association advocated for an RPS, and the California Public Utilities Commission discussed it as part of its restructuring decision.[57] Although California did not implement an RPS at that time (it did so in 2002), the clean energy advocacy community quickly picked up the idea.[58] A series of articles on the RPS appeared in the *Electricity Journal* in the late 1990s, some of them authored by UCS members.[59] Scientists associated with the UCS called for the adoption of an RPS because "renewables currently cost a little more than fossil fuels and, in a deregulated electricity market, could disappear, taking their many benefits with them."[60] In addition, since 1997 the UCS and other environmental groups have completed several influential studies that examine the costs and benefits of various national RPS proposals.[61]

The environmental groups' actions resulted in the introduction of numerous RPS proposals at the federal level. The Senate has passed a national RPS as part of comprehensive energy legislation three times since 2002. In 2007, for example, a large coalition including environmental organizations such as Greenpeace, the League of Conservation Voters, the National Audubon Society, the National Environmental Trust, the NRDC, Public Citizen, the Sierra Club, and US PIRG introduced a renewable electricity standard bill, which required utilities to acquire 20 percent of their electricity from clean, renewable energy by 2020. To build support for a national RPS, the coalition published reports that emphasized not only the bill's environmental benefits (reducing global warming pollution by an amount equivalent to taking 36.4 million cars off the road) but also its economic benefits (creating 185,000 new jobs from renewable energy development, and bringing $25 billion in income to farmers, ranchers, and rural landowners).[62] However, despite environmental organizations' extensive lobbying and constant support from some elected officials such as

Senator Jeff Bingaman (D-N.M.), this proposal failed to become law because of powerful opposition from major electric utilities and from influential politicians with links to the oil and coal industries.[63]

To build support for a national RPS policy, environmental groups cooperated not only with renewable energy associations but also with a number of electric utilities and energy developers. As one Sierra Club organizer argues, some utilities have an incentive to cooperate with environmental groups: "PG&E, which is a utility from California, has been supportive of an RPS, and also energy developers, like Florida Power and Light, were very supportive. Excel Energy has been very supportive of an RPS in Colorado and Minnesota and has not opposed them nationally. That's because a lot of those companies have already been making big investments in wind or solar, so they actually stand to benefit from any sort of future trading system of renewable energy credits."[64] Other utilities, however, strongly oppose any RPS policy. According to one organizer from Environment America:

> For the RPS a lot of the opposition has come from large utilities that don't want to be required to do anything that they are not doing already—groups like the Southern Company, which have really fought us pretty emphatically. A lot of the senators and representatives are not voting on the renewable electricity standard because these utilities have built up this mantra that the Southeast can't provide for itself through renewable electricity and that they would be unfairly prejudiced against it.[65]

As in the case of the PTC policy, the United States' failure to adopt a national RPS policy can be attributed to significant opposition from the oil, gas, and coal industries, as well as to an unfavorable political context. Significant opposition to a national RPS policy came from major electric utilities, particularly those located in the coal-rich southeastern United States. For example, one Sierra Club organizer argues:

> The biggest opposition we saw in trying to pass this legislation came from the utility industry. There are a small number of really powerful utilities that rallied together and ran ads almost every day in the Capitol Hill newspapers. They lobbied really aggressively against the RPS, saying it would increase energy prices and would put people out of business and that sort of thing. That was definitively our biggest opposition. Most of these utilities come from the Southeast; some of the big ones are American Electric Power, which was testifying at hearings, saying how much it was going to cost. Duke Energy is another big one, which is interesting because it came out in support of doing something on global warming, but when it came to the RPS, a specific policy that could cut global warming pollution, they were vehemently opposed. And then there is the Southern Company, another big company that was vehemently opposed. They have a big lobbying presence; there were many occasions where I would be meeting with a member of Congress and they would just have sat

down with a lobbyist from Duke Energy or from one of the other opposing utilities.[66]

Although U.S. environmental groups failed in their efforts to introduce a national RPS, they were successful in a number of states. Environmental groups introduced the concept of an RPS in California in the mid-1990s and lobbied hard for its adoption. The California RPS was created in 2002 under Senate Bill 1078 and further accelerated in 2006 under Senate Bill 107. These bills stipulate that California electricity corporations must expand their renewable portfolio by 1 percent each year until they reach 20 percent in 2010. But environmental groups pushed for even more ambitious goals. For example, the NRDC and other environmental groups formed a coalition and called on California's lawmakers to enact legislation establishing a 33 percent target for power from renewables by 2020. Noticing that California had installed less capacity from wind and other renewable energy sources than states such as Texas, Iowa, and Minnesota, the environmental coalition argued that a 33 percent target by 2020 was needed to reduce global warming pollution by more than 20 million metric tons and meet the global warming pollution cap by 2020.[67] The coalition was ultimately victorious: at the end of 2008, Governor Arnold Schwarzenegger signed an executive order that mandated an RPS of 33 percent by 2020, in addition to the mandated 20 percent by 2010.

Environmental organizations also pushed for the adoption of a Texas RPS, originally created by Senate Bill 7 in 1999. The Texas RPS mandated that utility companies jointly create 2 GW from new renewable energy sources by 2009. In 2005, a new senate bill increased the state's RPS requirement to 5.88 GW by 2015, and set a goal of 10 GW of renewable energy capacity for 2025. Because of this ambitious RPS, Texas installed more wind power than the rest of the country in 2001 alone (912 MW), and it became the number one wind energy producer in the nation in 2006. Environmental Defense was one of the main environmental organizations that shaped the Texas RPS. Its role was to design the initial RPS, negotiate with utilities, and lobby for subsequent revisions. According to one Environmental Defense organizer,

> We conceived the idea, we negotiated the deal, and we worked with the sponsor Steve Woolens of the House of Representatives. A deal was that the electric utilities can get the restructure they wanted, but the price to get that was that they had to agree to two things. Number one is they had to agree to clean up some of their old plants; and two, they had to agree to this renewable portfolio standard. Then, we were instrumental in drafting the regulations and we were instrumental in getting the amount raised in 2006 as well. We are working right now to get it raised to 10,000 MW.[68]

The success of RPS policies in Texas resulted not only from environmental groups' skillful negotiation with utilities, but also from the presence of a broad wind advocacy coalition. Although environmental protection was the main motivation for the adoption of the RPS, the installed capacity from wind has surpassed initial expectations because environmentalists formed a large and unusual alliance, which included farmers and conservative politicians. One Environmental Defense organizer emphasized the key role of farmers and conservative Republicans: "Initially the main driving force was the environment. It continues to be a driving force, but the fact that we required 2,000 MW by 2009 and then we more than doubled that in 2006, is a sign of two things: one, the technology is working and it is less costly than some predicted; two, we've gotten support from unusual people, like rural Republicans. We would not have been able to increase it that quickly if we didn't have this broad bipartisan support for wind power."[69] He went on to explain why farmers and local politicians like wind turbines:

> Wind turbines have a very small footprint. Farmers get paid a lot of money on top of their regular operations. The other thing is, of course, there are some affiliated jobs, everything from construction to maintenance. What we have done is we had mayors from the cities that are in the area come in and say what a great thing it has been for their community. That's why the sponsors of the RPS increase in 2006 were conservative Republicans from west Texas, people who are not typically big advocates for the environment.

Developing the right policy framework in Texas was a long process that took significant coalition-building efforts. One representative of the Texas Renewable Energy Industries Association (TREIA) sums up this argument:

> To the rest of the country this appears to have happened suddenly and overnight. Frankly, in many parts of the country they just can't believe or accept the possibility that Texas would become a leader in renewables. But we have been working on this since the mid-1970s. It is an incremental process that has laid the groundwork for the concept that renewable energy is simply one more piece of the broader energy industry and then that it will function in context with the rest of the energy industry.[70]

Energy experts also stress the role played by the conditions specific to the local energy sector. As the TREIA representative argues:

> We went through a period of time when natural gas was virtually eliminated from use in the generation of electricity in the early 1980s. That forced the use of coal in the state—those were federal policies that had to do with perceived shortages of natural gas at that time. That left us with a diversified resource base for electric generation, but at the same time it increased the pollution issue. We ended up with a more balanced set of resources over time, as natural gas became more abundant and cheaper. Once it was clear to the decision makers that you couldn't become dependent upon one resource and expect to survive and grow, it became more apparent that the more diversification we had, the better, and that played to the benefit of renewables. And then there is

the whole issue of economic development and the benefits that wind brings to the table. Rural areas of the state needing new ways to keep people employed. Renewable energy plays well in that issue: so, that was a factor. And then, there was the plain "We're not going to become a backwater for energy after having been an energy state for all this time" argument. If we are going to retain this, we have to pay attention and be on the forefront.[71]

Recently, U.S. environmental groups have also pushed for the introduction of FITs. Between 2007 and 2008, six states introduced FIT-type legislation: California, Illinois, Hawaii, Michigan, Minnesota, and Rhode Island. One of the environmental organizations that had a major influence on the introduction of FITs in the United States is the World Future Council (WFC). This transnational nonprofit organization was launched in 2004 on the initiative of Jakob von Uexkull, founder of the Alternative Nobel Prize, patron of FOE, and board member of Greenpeace Germany.[72] The WFC's position has been that the best way to tackle the climate change crisis is to adopt FITs as fast and widely as possible. It has argued that FITs have multiple advantages: they reduce carbon dioxide emissions, create jobs, secure the domestic energy supply, guarantee investment security, drive technological innovation, and provide fair market conditions. As the organization puts it, "We have to move quickly from our destructive, wasteful, and unfair use of fossil fuels to a new model where the production, distribution, and control of energy is clean, efficient, and affordable for everyone. One solution that has proven to help this renaissance is called a Feed-In Tariff (FIT), but we think it should be called 'the world's best renewable energy law.'"[73]

To promote the adoption of FITs, the WFC launched the Power to the People campaign in the United States, which attempts to revolutionize the U.S. energy sector. As the campaign organizers argue:

> Until now, energy has been controlled by a small number of large corporations. And governments have been giving about US$300 billion of subsidies to these corporations every year. Thanks to this arrangement, some people have become very rich. But most people are simply dependent on these companies for the power they need—and two billion people, or one third of the world's population, are still without access to reliable energy. **We need to switch.** With existing technology, we can capture enough renewable energy from the sun, wind, water, and the earth to power the world six times over. This technology can bring clean energy to everyone, everywhere. All we need is the political will and determined action to make it happen.[74]

Consequently, the campaign calls on citizens, business leaders, and legislators to support FITs. To build support for renewable energy FIT policies, WFC organizers have sent information to 4,500 decision makers in the United States, including state governors, energy legislators, and 700 Kyoto Protocol–supporting mayors. Additionally, the WFC has published research papers, organized workshops, and lobbied for the adoption of FITs in many other states.

Another organization that has been instrumental for the introduction of FITs in North America is the Alliance for Renewable Energy (ARE). The Alliance for Renewable Energy formed in early 2008 with the explicit mission of promoting renewable energy payments (another name for FITs) legislation in states and provinces throughout the United States and Canada. As Lois Barber and Paul Gipe, cofounders and cochairs of ARE, state, "Renewable energy payments [REPs] have proven to be the most widespread and effective legislation for the promotion of renewables. Our mission is to bring REPs to North America where they can help to rapidly increase our shift from fossil fuels to renewables, and in doing so improve our energy security and generate hundreds of thousands of new manufacturing jobs."[75]

In addition to transnational and national organizations such as the WFC and ARE, a number of local U.S. environmental groups have also supported the adoption of FITs. For example, in Wisconsin an organization called RENEW Wisconsin—a network of environmentalists, clean energy businesses, utility managers, and farmers—filed testimony with the Wisconsin Public Service Commission in 2006 calling for implementation of FITs in the state by 2008. In California, the Sierra Club published a report for the California Energy Commission in 2008 in which it declared its support not only for the adoption of a 33 percent RPS goal but also for "the implementation of a FIT for renewable projects, which could be modeled on the most successful FIT programs that have achieved renewable energy goals in Germany, Spain and France." The report went on to argue:

> While federal tax credits have built most of the wind power in the U.S., there have been frequent lapses in the credit. This has led to a "boom-and-bust" cycle in the wind industry that has stifled the growth of domestic manufacturing capacity. In this uncertain policy climate, investors are hesitant to commit money to manufacturing capacity, and wind farm developers have difficulty growing their business. Well designed FITs would allow the state to take charge of its own incentive structure for renewable energy without subjecting developers to the risks of arbitrary federal tax policy.[76]

The campaign for the introduction of FITs in North America has been spearheaded not only by environmental organizations but also by individual activists. One of the most active campaigners for FITs is Paul Gipe. He has been tirelessly promoting FITs by organizing workshops, giving presentations, and disseminating information through his personal website, appropriately named Wind-Works.[77] He is the main person responsible for the flurry of activity on FITs in California since 2006. For example, Gipe was one of the organizers and presenters at workshops on FITs held in Sacramento in 2007 and 2008, and his presentations convinced a number of members of the California Energy Commission to consider a renewable energy FIT modeled after the German and Spanish tariffs. His report for the commission's workshop on FITs argued emphatically, "The status quo is unacceptable. The situation calls for action and not endless discussion. The situation also calls for a full program, like those in Germany, France, and Spain that is implemented immediately."[78]

The outcome of the campaign for the adoption and implementation of renewable energy FIT policies in the United States will be decided not only by environmentalists' ability to mobilize effectively but also by the shifting political context. Environmental activists are aware that they are outgunned in their battle with large utilities and the fossil-fuel lobby. For example, the major electric utilities in California have opposed any change to the RPS program, arguing that "there would be nothing to gain by switching to a feed law except higher prices."[79] But, as Paul Gipe argues, "there's no greater endorsement of proposed renewable energy policy than the knee-jerk opposition of the state's two largest electric utilities."[80] Indeed, there are some signs that environmental activists are encountering a more-favorable political context, both at state and federal levels. For example, Congressman Jay Inslee (D-Wash.) recently declared his strong support for renewable energy and introduced FIT legislation in Congress under the title Renewable Energy Jobs and Security Act.[81]

It is worth emphasizing that the social and political context has been less favorable in the United States than in many European countries. The United States has a very powerful fossil-fuel industry that has many political allies and the ability to shape the mass media's framing of global climate change by buying a large amount of advertising space.[82] No other economically advanced country, for example, has top politicians who recently stated that global climate change is a "hoax" or "an article of religious faith."[83] No other economically developed country has a mass media that has presented a biased view of global warming and routinely stated that global climate change is controversial and theoretical. For example, Dispensa and Brulle (2003) show that, in contrast to scientific journals and to media in other countries, the U.S. media presents a biased view of global warming by systematically including the opinions of a small minority of global warming dissenters. Similarly, Boykoff and Boykoff (2004: 125) show that the U.S. prestige press's—meaning the New York Times, the Washington Post, the Los Angeles Times, and the Wall Street Journal—adherence to balance "actually leads to biased coverage of both anthropogenic contributions to global warming and resultant action." And in no other industrialized country has the public supported President George W. Bush's decision to reject the Kyoto Protocol in 2001 as much as in the United States.[84] The fossil-fuel lobby continues to influence the public debates on global climate change in the United States and has intensified its efforts to prevent the adoption of meaningful legislation to address climate change.[85]

A more-favorable political context, however, emerged after the election of President Barack Obama, who made renewable energy a priority of his energy policy. Some positive signs could be seen soon after his election; for example, the U.S. Department of Energy announced in May 2009 that it planned to provide $93 million from the American Recovery and Reinvestment Act to support the further development of wind energy in the United States.[86] In 2009, the U.S. wind industry also broke all previous records by installing 9.922 GW of new generating capacity. The catalyst for this record-breaking year was incentives from the American Recovery and Reinvestment Act; in fact, before the passage of the American Recovery and Reinvestment Act, wind industry experts anticipated that in 2009 wind power development might drop by as much as 50 percent from 2008 levels.[87]

The Environmental Movement and Renewable Energy Policies in Canada

Canadian provinces began adopting RPS policies only in 2000, when British Columbia adopted the first voluntary RPS. In 2004, four Canadian provinces had voluntary RPS policies: British Columbia, Alberta, Ontario, and Nova Scotia. In 2008, New Brunswick, and Prince Edward Island also adopted RPS policies, while Quebec and Manitoba adopted a similar policy that mandated a capacity increase for wind power (4 GW and 1 GW of wind power respectively by 2015). Additionally, in 2006 Ontario became the first region in North America to adopt a FIT.[88]

It is not possible to fully understand why Canada adopted these energy policies without examining the campaigns against nuclear power and climate change. Mobilizations against nuclear energy in Canada began somewhat timidly in the 1970s. Because criticism of civilian nuclear power came mostly from the United States, it was initially dismissed by the Canadian nuclear industry on the grounds that Canadian reactors—known as CANDU reactors—were "substantially different from, and also safer than, American designs" (Mehta 2005, 39). The antinuclear campaign was initially active mostly at the local level: the periodic relicensing of nuclear power plants by the Atomic Energy Control Board often provided an opportunity for local groups to organize local protests. In Ontario, groups such as CANTDU—a pun on "CANDU"—and Durham Nuclear Awareness have fought against relicensing of local nuclear power plants since the 1970s and 1980s. In Nova Scotia, New Brunswick, and Prince Edward Island, the Maritime Energy Coalition not only opposed the construction of nuclear power plants but also promoted renewable energy (Mehta 2005, 41). In Quebec, a group named SVP called for a moratorium on new nuclear power plants in 1977 and built a coalition of local associations asking for the democratization of the decision-making process on energy policy (Babin 1985, 177).

The antinuclear campaign grew stronger with the formation of the Canadian Coalition for Nuclear Responsibility (CCNR) in 1975 by forty-five environmental groups; before the end of the decade, CCNR had grown to over two hundred groups (Babin 1985, 159). Criticism of the Canadian government's plan to expand the use of nuclear power became louder toward the end of the decade when CCNR-affiliated scientists began attacking the reliability and safety of the CANDU reactors and launched a national petition drive demanding that the federal government set up an independent commission of inquiry into all aspects of the nuclear program. These efforts attracted the attention of a number of politicians from across the political spectrum; some federal and provincial politicians lent their support to antinuclear activists, and a few called for moratoriums on nuclear power (Babin 1985). The Canadian antinuclear campaign, however, was never strong enough and did not have enough influential allies to impose a moratorium. Although nuclear energy production declined at the end of the 1990s because of technological problems, it grew again at the beginning of the twenty-first century due to improved reactor performance and refurbishment.[89]

Starting in the 1990s, many Canadian environmental groups involved in the antinuclear campaign also became involved in campaigning against climate change. The Campaign for Nuclear Phaseout (CNP), for example, is an organization that combines the energy of antinuclear and anti–climate change organizers. It is a coalition of major organizations—such as Greenpeace and the Sierra Club—and numerous grassroots environmental groups. The CNP is headquartered in Ottawa, Ontario, and lobbies the federal and provincial governments to phase out nuclear power, increase energy efficiency, and promote renewable sources of energy such as wind power.[90] In 2003, the CNP commissioned a report titled "Phasing Out Nuclear Power in Canada," which argued that it was possible to phase out not only nuclear power but also coal power plants in some provinces by increasing energy efficiency and deploying renewable energy. As the report stated, "Our choice need not be between nuclear power and coal; it can be instead a choice between the unsustainable energy options based on nuclear and coal, and more sustainable options based on energy conservation, efficiency improvements, cogeneration, renewables and other alternatives. Seen in this light, the decline of the Canadian nuclear program presents an opportunity for an orderly transition to a more sustainable electricity future."[91]

Environmental groups working against nuclear power and in support of clean energy exist in many Canadian provinces. In Alberta, the groups Citizens Advocating the Use of Sustainable Energy (CAUSE) and Coalition for a Nuclear Free Alberta oppose plans for nuclear development and support wind and solar power. In Ontario, the Pembina Institute, the World Wildlife Fund, the David Suzuki Foundation, Greenpeace, and the Sierra Club joined forces to promote a plan called Renewable is Doable, which seeks to identify electricity scenarios that will meet future power demands for the province without the use of nuclear power and coal. At the federal level, environmental organizations have not only built capacity for lobbying officials but also worked on producing solid research in an attempt to shape the energy policymaking process. The David Suzuki Foundation, for example, has produced numerous high-quality reports on climate change such as "Power Shift: Cool Solutions to Global Warming." As a result, one study concluded that the Canadian environmental movement "had considerable success in pushing issues like species extinction and climate change onto public and government agendas" (Wilson 2002, 50).

The environmental movement has been more successful in Canada than in the United States in pushing the climate change issue to the top of the political agenda. Indeed, while the United States has never ratified the Kyoto Protocol, Canada ratified it in 2002. However, the Canadian environmental movement has not succeeded in pushing the federal government to adopt a coherent renewable energy policy. Three main reasons account for this failure. First, during the 1970s and 1980s the Canadian environmental movement did not create grassroots organizations or institutes that focused on developing policy innovations in the field of renewable energy, as the German and Danish movements did. Second, domestic environmental organizations did not lobby very hard for a federal RPS policy during the 1980s or 1990s because Canada has a relatively clean energy sector: at the beginning of the twenty-first century, the country produced about 61 percent of its electricity from hydropower, and in terms of hydropower installed capacity, it was number two in the world. Third, Canada has strong fossil-fuel and nuclear energy lobbies—for example, Canada ranks fourth in the world in terms of coal reserves, second in the world in terms of oil reserves, and second in the world in terms of uranium reserves.[92]

While the environmental movement has had a minimal impact on federal energy policy in Canada, it has had a significant influence on regional energy policies. Ontario is perhaps the region where this influence has been the largest. In 2004, the new provincial government adopted an RPS policy that required that Ontario produce 5 percent of its electricity from renewables by 2007, and 10 percent by 2010.[93] Moreover and as noted above, in 2006 Ontario adopted the first FIT in North America. Consequently, Ontario was producing over 1.1 GW from wind power by mid-2009, or approximately 41 percent of the total installed capacity in Canada.[94]

The Ontario FIT has similarities to many European tariffs: it has prices that differ between technologies, it provides for simplified interconnection with the grid, it is reviewed every two years, it offers long-term contracts of twenty years, and it limits project size to encourage distributed generation. The implementation of the program started in 2006 for qualified producers of up to 10 MW of power, but was expanded in 2007 to include larger-scale producers. Perhaps the most interesting thing about Ontario's FIT—called the Standard Offer Program—is that it resulted mainly from the activism of one person: Paul Gipe.

It is very likely that Ontario would have had no FIT if Paul Gipe had not decided to move to Canada. His decision resulted from his perception that this province offered an opportunity that did not exist anywhere else in North America; as he explains,

> I've been a proponent of feed-in tariffs since the '90s, and publicly since 1998. In 1998, in campaigning for a seat on the board of directors of the Wind Energy Association, my campaign statement said specifically that it was time for this association to abandon its failed attempt at extending the tax credits and should instead work for a much more productive and beneficial system that would allow everyone to develop wind energy—and that is the system of feed-in tariffs that was being used in Germany. Since then my knowledge that this is the premiere policy mechanism for the rapid development of massive amounts of renewable energy has just grown. In 2003 I decided again, as I did earlier in my career, to put my career where my mouth was. Rather than just being an advocate, writing articles and saying that the Germans are doing great things and we should do the same, in 2003 I said I should do this, no one else is, so I will try to make that happen. And, of course, I couldn't make it happen in the U.S., so I went to Canada.[95]

Paul Gipe made the decision to promote a renewable energy FIT policy in Ontario because this province had numerous proponents of community-owned wind farms. As he recollects,

> I'm well-known for promoting community wind power, and there's a group in Ontario that supports community wind. When they dedicated their first community-owned wind turbine, a cooperatively owned wind turbine in Toronto, they invited me up for the dedication. After that, in 2003 I was looking for work and this group called me asking me to help them locate a temporary executive director. I said, "Sure, I'll circulate it for you but I want to apply for the job." I convinced their board of directors that they needed to pursue the feed-in tariff policy. In fact it was a bit shocking to me when they said, "You know what? You're right, this is exactly what we need; can you start in February?" They're great people and the politics in Toronto were just right, I was in the right place at the right time. I did the right things, and we had the first North American policy. Not perfect, but it's the first.[96]

But Paul Gipe was not alone: a number of environmental activists and renewable energy advocates from the Ontario Sustainable Energy Association (OSEA) worked hard for the adoption of the Standard Offer Program. The OSEA was created in 2003 by Canadian and U.S. environmental activists who wanted to take advantage of the new political context in Ontario in order to "facilitate Ontario's transition to a sustainable energy economy based on 100 percent renewable power."[97] At that time, the Liberal Party came to power on a strong environmental platform, which included closing

Ontario's dirtiest coal-fired power plants. Paul Gipe and other OSEA members started a campaign for the adoption of a FIT in Ontario in early 2004.

By the end of 2004, OSEA had been commissioned by the Ontario Ministry of Energy to propose a policy for developing community-owned renewable power projects in the province. In 2005, OSEA published a report arguing that FITs were ideal for Ontario because they would "unleash the entrepreneurial spirit of Ontarians, provide more renewably-generated electricity, more economic activity, and more jobs in the manufacturing of wind turbines and solar panels than any other means available to the province."[98] The report emphasized both the environmental and economic benefits of renewable energy FIT policies; for example, it estimated that if just half of Ontario's fifty-five thousand farmers installed one average-size wind turbine, they could pump C$4 billion through the rural Ontario economy. In 2008, OSEA and other environmental organizations launched the Ontario Green Energy Act campaign to revise the Standard Offer Program and "provide a roadmap for a renewable energy future with a vibrant renewable energy economy and a culture of conservation."[99]

The Standard Offer Program of 2006 and its subsequent strengthening resulted directly from the work of environmental activists involved in OSEA and other NGOs. It also resulted from the fact that, since 2003, these organizations had encountered a political context favorable to environmental activism. Indeed, two facts show that FIT proponents had important allies among elected officials in Ontario. One is that Ontario's Liberal Party adopted a resolution supporting FITs at its conference on energy policy at the end of 2004. Although the resolution was nonbinding, it was a major event since it was the first time a major political party in North America had declared its support for FITs.[100] The other fact is that OSEA received significant financial support from two governmental agencies: the Ontario Ministry of Agriculture, Food and Rural Affairs; and the Ontario Ministry of Energy and Infrastructure.

Conclusion

This chapter has examined the way in which environmental groups and activists have shaped the energy policymaking processes in countries that have very good wind potential but a social context that is less favorable to the environmental movement. It has shown that the environmental movement can contribute to the adoption and implementation of policies such as an RPS through its campaigns against nuclear power, air pollution, and global climate change. However, while environmental groups mobilize large green-energy advocacy coalitions to shape the adoption and implementation of pro–renewable energy policies, their ability to reach their goals is severely limited when they lack influential political allies, and when they face a biased mass media and less-favorable public opinion.

Similarly to the environmental movements analyzed in the previous chapter, the environmental movements in the United Kingdom, United States, and Canada started to mobilize against governmental plans to build nuclear power plants in response to the energy crisis of the 1970s. In contrast to the groups studied in the previous chapter, however, the antinuclear mobilizations in these three countries were less influential. In the United Kingdom, the antinuclear mobilizations were weak and coopted by the government. Only with the recent ascent of global climate change to the top of the environmental agenda and a gradual opening of the political opportunity structure have environmental groups been able to influence the revision of the RO policy and the adoption of a FIT.

In the United States, the antinuclear mobilizations contributed to the emergence of research institutes that promoted alternative energy but were not successful in promoting a moratorium on nuclear power. Moreover, global climate change has been a very controversial issue, and U.S. environmental groups' attempts to influence governmental policies on this issue have been opposed by numerous electric utilities and fossil-fuel industries. Consequently, environmental groups' attempts to build large renewable energy advocacy coalitions to promote policies such as an RPS, PTCs, and FITs have had relatively little success at the federal level. These efforts, however, have been more successful in states that adopted and implemented an RPS and other renewable energy policies. Likewise, in Canada environmental groups had little impact on the federal policymaking process but contributed to some provincial governments' decisions to adopt RPS and renewable energy FIT policies.

Chapter 2 and this chapter showed that environmental groups may have a top-down influence on the wind energy industry, by shaping the energy policymaking process. Indeed, environmental groups have made an essential contribution not only to the invention of renewable energy FIT and RPS policies, but also to their adoption and implementation. The next chapter will show that environmental activists and organizations may also have a bottom-up influence on the industry, by creating demand for renewable energy and contributing to the emergence of a voluntary green-power market.

4

From Thinking Globally about Climate Change to Acting Locally on the Energy Challenge

When in the course of human events, a nation's energy policies compromise the health, security and prosperity of its people, and cause global climate disruption, a new path must be taken. We, the youth of the United States of America, declare our independence from dirty energy. We demand that our nation reject dirty energy sources such as fossil fuels, nuclear and incineration, and make a strong commitment to energy efficiency and clean, renewable energy technologies such as wind and solar.

—Energy Action Coalition, http://www.energyaction.net/documents/declaration.pdf (accessed December 2007)

Creating Consumer Demand for Wind Energy

In 1998, the New Belgium Brewery, located in Fort Collins, Colorado, took an employee vote and became the first brewery in the United States to subscribe to wind-powered electricity. Other breweries soon followed its example: the Uinta Brewing Company decided to switch to 100 percent wind-generated electricity in 2002, and the Brooklyn Brewery made a similar decision in 2003, announcing "There's wind in our ales...Here at the Brewery we make use of alternative energy because we truly care about our environment and community."[1]

Breweries, however, are not the only companies that have switched to green power. Whole Foods Market, the largest natural and organic foods supermarket in the United States, announced in 2006 that it will purchase wind power and other renewable energy certificates (RECs) to offset 100 percent of the electricity used in all of its stores. PepsiCo, one of the world's largest food and beverage companies, announced in 2007 that it will purchase enough RECs (mostly from wind) to match the purchased electricity used by all PepsiCo U.S.-based manufacturing facilities, headquarters, distribution centers, and regional offices—a purchase estimated by the U.S. Environmental Protection Agency to be the same amount of electricity needed to power nearly ninety thousand average American homes annually.[2] Even Walmart, a corporation that has not been known for being an environmental leader, announced at the end of 2008 that it will purchase wind-generated electricity to power up to 15 percent of its 360 stores and facilities in Texas.[3]

A growing number of universities have also "declared independence from dirty energy" and demanded that the United States reject dirty energy sources and make a strong commitment to clean, renewable energy technologies. Most of these universities are buying wind energy certificates for a significant percentage of their electricity consumption, and some have even installed their own wind turbines. At the same time, more and more local governments and individual electricity customers are purchasing wind power.

These voluntary decisions to purchase RECs have had a major impact on the U.S. market for renewable energy and, in particular, on the wind power market. The National Renewable Energy Laboratory (NREL) estimates that over 10.6 GW of new renewable energy capacity was installed in the United States between 1997 and early 2007. Much of this new renewable energy capacity would have not been possible without consumers' voluntary decisions to purchase electricity supplied from renewable energy sources. Voluntary green-power markets provided support for over 3.1 GW of "new" renewable energy capacity additions in this period, most of the remaining renewable energy generation from recent capacity additions being used for compliance with various policy mandates such as a renewable portfolio standard (RPS). Since wind energy provided over 62 percent of green-power sales, customers' voluntary decisions to pay extra for renewable energy were responsible for the installation of almost 2 GW of wind power capacity between 1997 and 2007. Therefore, as NREL energy analysts have concluded, green-power purchases provide support for a significant fraction of new renewable energy and wind power projects at the national level.[4]

What accounts for the "rising tide" of individuals, companies, universities, and local governments that choose wind power and other forms of renewable energy?[5] Why has consumer demand for wind energy increased over the last few years, particularly in countries such as the United States? This chapter shows that, because American environmental groups have had very little success in influencing federal energy policies, many environmental activists and organizations have been organizing campaigns focused on creating local demand for renewable energy. It builds on studies of social movements' influence on organizations, and focuses on the environmental movement's contribution to decisions by U.S. colleges, universities, and corporations to purchase wind energy.

The chapter examines a number of orienting research questions. Existing research suggests that the environmental movement may pressure organizations to change "from the outside" by organizing campaigns that use different tactics: petitions, lawsuits, boycotts, protests, and shareholder activism.[6] Studies also suggest that the environmental movement may contribute even more to change "from the inside" by changing the ideological commitment of organizational members and turning them into "environmental mediators"— individuals who are members of the environmental movement and also professional members of an organization or institution.[7] Therefore, the chapter addresses three exploratory questions: How do environmental organizations influence university students' and administrators' commitment to address climate change? How do they influence company employees' commitment and capacity to address climate change? And what is the role of protests and other forms of collective action against climate change in the growth of the voluntary market for wind power?

It is important to note that the chapter does not examine how the environmental movement creates consumer demand for renewable energy at the household or individual level because of the lack of data, and not because this does not matter. Much of the literature on determinants of individual pro-environment behaviors such as energy conservation focuses on demographics, knowledge and information, political attitudes, and values such as postmaterialism or egalitarianism.[8] This literature suggests, however, that the environmental movement may create household demand for green energy by changing individuals' values as well as by increasing the salience of their "environment identity."[9] Environmental groups also generate this type of demand by disseminating information and even by selling green power to their members—for example, Greenpeace Germany has been a retailer of green power since 1999.[10]

Campuses Declare Independence from Dirty Energy

Between 2000 and 2008, more than fifty colleges and universities in the United States purchased RECs, and most of those RECs were purchased from wind power developers. A small number of colleges and universities also installed wind turbines on their campuses, and many others were planning to purchase wind power from independent producers or to develop their own wind farms. Table 4.1 shows a selective list of colleges and universities—only those that are nonreligious and offer at least a bachelor's degree—that were obtaining a significant percentage of their electricity from wind power and other renewables by March 2008.

Environmental groups contribute to college and university decisions to purchase wind power primarily by raising awareness of climate change and other environmental problems among students, faculty, and administrators. Colleges and universities situated in states with a high density of

Table 4.1. Colleges and Universities That Were Obtaining a Significant Percentage of Their Electricity from Wind Power and Other Renewables by March 2008

College or University	Source
American University	Wind
Bates College	Wind
Bowdoin College	Various
California State University	Various
Carleton College*	Various
Carnegie Mellon University	Wind
Clemson University	Various
Colby College	Wind
College of the Atlantic	Wind
Colorado State University*	Wind
Connecticut College	Wind
Drexel University	Wind
Duke University	Wind
Evergreen State College	Wind
Green Mountain College	Wind
Hamilton College	Various
Harvard University	Various
Lewis and Clark	Various
Massachusetts Maritime Academy*	Wind
New York University	Wind
Northwestern University	Wind
Oberlin College	Various
Oregon State University	Various
Pennsylvania State University	Wind
South Dakota State University	Wind
State University of New York at Buffalo	Wind
Syracuse University	Wind
Texas A & M University	Wind
University of California	Various
University of California at Santa Cruz	Wind
University of Colorado at Denver	Wind
University of Denver	Wind
University of Iowa	Various
University of Massachusetts at Amherst*	Wind
University of Massachusetts at Lowell	Various
University of Minnesota at Morris*	Wind
University of Oregon	Various
University of Pennsylvania	Wind
University of South Dakota	Wind
University of Southern Maine	Various
University of Utah	Wind
University of Washington	Various
University of Wisconsin at Green Bay	Wind
University of Wisconsin at Oshkosh	Wind

(*continued*)

Table 4.1.   (*Continued*)

College or University	Source
Warren Wilson College	Wind
Wesleyan University	Wind
Western Washington University	Wind
Whitman College*	Wind
Yale University	Various

Source: U.S. Department of Energy, The Green Power Network; U.S. Environmental Protection
Agency, Green Power Partnership; Association for the Advancement of Sustainability in Higher
Education, Green Power on Campus.

* University or college owns wind project; all other institutions purchased renewable energy
certificates (RECs).

environmental groups are likely to conserve energy and purchase wind
power because their students and administrators have higher-than-average
levels of knowledge and concern about global climate change. This fact is
reflected in a comment by a California State University administrator:

> Say what you will about folks that live out here in "the land of the
> fruits and nuts," but we care about the environment and being envi-
> ronmental stewards. Our residential and commercial building stan-
> dards have really set the standard for the rest of the country; our per
> capita energy use has remained flat for the last thirty years. When you
> have statewide policies at that level, that impacts everybody at every
> aspect of their life, whether their home life or their work life. For the
> most part, I would say, there is a reasonable amount of general aware-
> ness of climate change and the environment; certainly, you can find a
> number of folks who are clueless, but, by and large, the number of
> environmentalists is higher than in the heartland of the United
> States.[11]

Yet, only some colleges and universities in environmentally progressive
states are buying wind energy; moreover, a number of colleges and universities in states that are not environmentally progressive are also buying wind
energy. What distinguishes those who buy wind power from those who do
not is the combination of two factors: well-organized student campaigns for
clean energy, and the presence of top-level administrators committed to
addressing climate change. As one student organizer observed,

> In order for the administration to be able to spend money on a renew-
> able energy purchase for the university, they need to have some sort of
> public support. Having an active student group to be able to show that
> broad base of support on campus will make them a lot more comfort-
> able and more willing to take that decision. At the same time, it's not
> going to happen at a school without the commitment from administra-
> tors at the top. A lot of times students have actually passed through

their student government referendums to increase their student fees, but it has actually been rejected by the administration or by the university's board of trustees. There have been instances in which the students were not allowed to tax themselves to pay for renewable energy because of a lack of support from the president or top-level administrators.[12]

The importance of having support from top-level administrators who are committed to addressing climate change was emphasized by one student involved in the campaign for clean energy at New York University. In her words:

We were lucky to have a lot of people at the top for whom climate change is a personal issue that they take very seriously. I know that our university president, when he announced the initiative, said he started to look into these things because he really worried about his grandchildren and what the world was going to be like for them. When our executive vice president came last year, this was something that he really cared about and he was one of the people that really pushed and moved us forward. The two vice presidents that are cochairs of the sustainability task force, both very influential people in the university, are also very concerned about this.[13]

But why did students organize campaigns for clean energy and why did college and university presidents commit to addressing global climate change? To understand this, it is necessary to examine the factors that contributed to the emergence of two campaigns: the student campaign for clean energy, and the campaign for college and university presidents' climate change commitment. Both of these campaigns can be seen as part of the environmental movement's broad campaign against global climate change. Yet, each of them have specific histories and require separate analyses.

The Student Campaigns for Clean Energy

While in the past U.S. students have mobilized on issues ranging from civil rights and apartheid to peace and social justice, currently the most significant student campaign is on the issue of clean energy.[14] One journalist observed that, for the new generation known as the Millennials, "climate change is emerging as the defining issue of their time, just as civil rights or Vietnam might have been for the generation before."[15] Another journalist noted:

In recent American history, college students marched through the Deep South during Freedom Summer or barnstormed New Hampshire on behalf of anti-establishment candidates such as Eugene McCarthy. But today, [some] students aren't spending their summer vacations effecting political change. Global warming is the issue that motivates them instead, driving them to work long hours for little pay.... Their mantra of "green is the new red, white, and blue" carries the same urgency that "we shall overcome" did decades ago.[16]

Indeed, when students at Swarthmore College announced the success of their campaign for wind power, they emphasized that their main motivation was to address global climate change in the face of governmental inaction. Their 2007 press release stated: "Our victory shows that, while there has been a lack of political will among our national leaders to adequately address global warming, institutions of higher learning, followed by state and local governments, can lead the way towards a clean energy future."[17] In addition, students have been motivated by the connection between the consumption of fossil fuels and various social problems. As one student activist remarked:

> We were motivated mainly by concern about climate change, but also by a lot of different issues that kind of centralize in the desire for clean energy. I think a lot of students have been concerned about the war, especially in 2003 and before that, a lot of students have recognized the links between oil resources and war. Also, students have been concerned about human rights issues, about poverty. There are a lot of different links between extraction of fossil fuels and the major social issues that concern college students.[18]

The growing awareness of the problems associated with the consumption of fossil fuels has resulted in numerous individual actions to reduce the personal carbon footprint. However, as one student environmental activist put it, the focus on individual solutions "rings hollow to a lot of people"; for many student activists the solution is "to organize and organize and organize."[19] Consequently, students on campuses from the East Coast to the West Coast and in between have mobilized to reduce consumption of fossil fuels and build support for renewable energy on campuses. To understand how the student campaigns for clean energy emerged and spread to campuses across the country, it is important to examine the role of local and national environmental organizations.

Greenpeace and Ozone Action are two of the organizations that played a key role in the emergence of the student campaigns for clean energy.[20] In anticipation of the sixth session of the United Nations Framework Convention on Climate Change Conference of the Parties (COP 6)—which was held in The Hague, Netherlands, in November 2000—these environmental organizations sponsored over 220 students from campuses around the United States to travel to The Hague and to participate in demonstrations outside the COP 6 meetings. The November 2000 demonstrations turned out to be some of the largest and most contentious mobilizations around the issue of global climate change. As the American delegation pushed for loopholes that prevented cuts in the use of fossil fuels, and negotiations reached a deadlock, demonstrators staged street protests, climbed buildings and unfurled banners, blockaded doorways, passed out Christmas stockings stuffed with coal, and even spattered the top American negotiator with a cream pie at a news conference.

Many of the American students who protested at The Hague came back energized and formed student groups that aimed to create change at the grassroots level. One of these groups was Kyoto Now! a group formed at Cornell University in 2001. Kyoto Now! aimed to commit Cornell to a goal set by the Kyoto Protocol: to reduce emissions of greenhouse gases by 7 percent below 1990 levels by 2010. Because the university administration initially resisted this request, students organized a sit-in and various demonstrations that eventually convinced the university to accept their demands. This resulted in the creation of the Kyoto Task Team and several projects that allowed Cornell to reach the Kyoto goal, mostly by increasing energy efficiency. Cornell students also started a campaign to build a wind farm, but because the project encountered strong NIMBY opposition from local groups, they later reoriented their efforts toward persuading the university to buy wind power certificates.

Another environmental organization that contributed resources to student campaigns for clean energy, particularly in Colorado, is Western Resource Advocates (WRA).[21] This organization played a key role in the decision to purchase wind energy at the University of Colorado at Boulder—the first such decision in the country. Western Resource Advocates has worked with Xcel Energy, a Colorado-based utility, to develop a program called Windsource in which customers could pay slightly more on their electric bills to receive electricity generated by Colorado's first wind farms. Beginning in 1997, WRA approached different people at the University of Colorado at Boulder to encourage the university to buy wind power, but the added cost was always a barrier. In 2000, WRA helped students with strategy and funding for advertising and organizing a campaign to get a fee increase to pay for wind energy. Due to WRA's help, the Clean Energy Now! campaign was very well organized and included various tactics such as hiring a research company to survey the level of student support for wind power, making posters, placing ads in the student paper, writing letters to the editor, distributing fliers, handing out pinwheels that symbolized wind turbines, and organizing educational events with well-known environmental activists such as Denis Hayes.[22] Consequently, their campaign was a huge success and students' votes showed overwhelming support for wind energy.

Still another regional organization that offered crucial resources to student campaigns for clean energy is the Southern Alliance for Clean Energy (SACE).[23] In 2003, SACE launched an initiative focusing on higher-education schools in the Southeast, with the goal of "helping colleges and universities in the Southeast become more sustainable in their daily practices through the creation of campaigns that would bring renewable energy to the campus community."[24] According to one SACE organizer, the organization contributes to student campaigns for clean energy mainly by distributing information and organizing regional and national conferences: "We're now connected to, I would say, seventy-five or so campuses in the Southeast in some way. There are different ways we [do] outreach. We have our own regional conference that I helped start a while ago. It's an annual regional conference. And then this fall we're having the first big national student youth and climate conference."[25]

The Sierra Club, a national environmental organization, has also contributed to the growth of student campaigns for clean energy around the country. By early 2004, the number of student groups that worked on clean energy issues had reached a critical mass. The first Fossil Fools Day was held with support from major environmental organizations on April 1, 2004, with the intention of focusing attention on the United States' addiction to fossil fuels. According to organizers from the Sierra Student Coalition—the student-run arm of the Sierra Club—this event was planned because "admitting you have a problem is always the first step in breaking an addiction. Luckily, President [George W.] Bush has taken that first step by admitting to the nation that America is addicted to oil. Unfortunately, Bush and scores of other politicians and corporate CEOs have stopped right there and are exhibiting the classic signs of an addict: denial, aggression, avoidance, and shifting the blame."[26]

The success of the first Fossil Fools Day inspired students from the Sierra Student Coalition and other environmental groups to create the Energy Action Coalition (EAC). The newly formed EAC planned an Energy Independence Day for October 14, 2004, an event that—with over 280 local actions—became one of the largest youth climate actions in the world. On this day, the EAC released its "Declaration of Independence from Dirty Energy" and sent signatures of support from more than twenty-seven thousand students nationwide to members of Congress and presidential candidates. As this document stated,

> The United States spends billions of dollars every year subsidizing powerful dirty energy corporations like ExxonMobil, Exelon Nuclear and Peabody Coal. Our nation must shift these investments into energy efficiency and a new generation of clean energy sources such as wind and solar, which will create millions of new jobs and improve our economy.... We, the youth of the United States of America, challenge all politicians and leaders of our institutions to lay out their plan for a complete transition beyond dirty energy. In defense of ours and future generations, we declare our firm commitment to a clean energy future![27]

The "Declaration of Independence from Dirty Energy," however, was only the beginning of the EAC's campaign for clean energy. In May 2005, EAC partners came together to discuss a nationwide campaign named the Campus Climate Challenge. The main goal of this campaign was to mobilize young people on college campuses across the United States and Canada in order to "win 100% Clean Energy policies at their schools."[28] In January 2007, the

Campus Climate Challenge campaign organized the largest youth mobilization in the history of the climate change movement to date, in a week of action that had events on nearly six hundred campuses in the United States and Canada and reached over fifty thousand students. And in April 2007, students from as many as 1,400 communities around the country came together in the first "open source, web-based day of action dedicated to stopping climate change," called Step It Up National Day of Climate Action, and held up banners with the message "Step It Up, Congress: Cut Carbon 80% by 2050."

This high level of student mobilization for clean energy would have been impossible without the support students received from environmental organizations. The campaign's national coordinating group, the EAC, has at its core large environmental groups that each have a student-run arm (groups such as the Sierra Club, Greenpeace, Friends of the Earth, and the National Wildlife Federation), as well as regional and local environmental groups—such as the Southern Alliance for Clean Energy, the Chesapeake Climate Action Network, and various chapters of Student Public Interest Research Groups (SPIRGs) and Student Environmental Action Coalitions (SEACs). One student activist describes how environmental organizations offered key resources to the student campaigns for clean energy:

> In the fall of 2003, a few innovators who worked with students across the country and in different regions came together and had the first national day of action for clean energy on college campuses. That was organized by Greenpeace, a group called the Climate Campaign, and the Student Environmental Action Coalition. It was organized all over the country. There were about sixty-five events that happened on college campuses and different types of advocacy events—spreading awareness, having a table to talk about wind energy, doing creative theater, all sorts of different things.[29]

In fact, the Climate Campaign was itself a network of ten national and regional student environmental networks: ConnPIRG, ECO-Northeast, EnviroCitizen, Free the Planet! Greenpeace Student Activist Network, MASSPIRG, NJPIRG, the Sierra Student Coalition, the Student Environmental Action Coalition, and SustainUs.[30]

Environmental organizations' resources made possible the formation of the EAC and the coordination of the national campaign for clean energy. According to the same student organizer quoted above,

> The following spring we had another day of action event about twice the size—130 campuses around the U.S. and extending into Canada. And that following June, in 2004, we had a coming together with a group of about twenty-two coalition partners to formally create the Energy Action Coalition. That led into a joint campaign and in mid-2004 it became a more united effort and we were getting the same types of resources to students. Then the unified effort for wind energy

really took off in the fall of 2006, when we launched a major united campaign called the Campus Climate Challenge. This was a joint campaign of about thirty organizations that are funded together to support and strengthen student campaigns for the reduction of greenhouse gas emissions and for clean energy. There are about fifty paid organizers who work with the challenge, work with students in different regions of the U.S. and Canada now. We have a unified strategy; we've developed a model policy that students can work with; we've got a tool kit to help them. There are trainings all the time; there are summits two or three times a year all over the country.[31]

The participation of major national and regional environmental groups also transformed the student campaigns on clean energy from volunteer-based, dispersed activism to well-coordinated, professional activism. For example, the main organizer of the Climate Campaign was Billy Parish, a former Yale student who was the head of the Yale Student Environmental Coalition.[32] Working initially as a volunteer, in 2004 Billy Parish obtained grant funding for the Climate Campaign from the Kendall Foundation through Clean Air-Cool Planet, an environmental organization that partners with campuses, communities, and companies throughout the Northeast to help reduce their carbon emissions.[33] Indeed, the involvement of professional organizations such as Clean Air-Cool Planet, Greenpeace, and the Sierra Student Coalition attracted major funding from private foundations and allowed the EAC to hire almost fifty paid organizers.[34]

Finally, it is important to emphasize that student environmental groups have also played a part in the emergence of the campaigns for clean energy on college campuses. Indeed, many of the students and administrators inter-viewed declared that their local campaign benefited tremendously from the involvement of one or more environmental groups on campus. In some cases, the environmental groups behind the campaigns were old and experienced—as in the case of the Environmental Center at the University of Colorado at Boulder, the first student-operated environmental center in the country.[35] In other cases, these groups were new—as in the case of Green Arch, a student organization founded in 2005 and dedicated to promoting sustainability at New York University.

Student environmental organizations may play two important parts in persuading university administrators to purchase wind power certificates. First, they may provide material resources by promoting energy conservation measures on campus and using the savings to buy renewable energy. For example, in 2002 students from Duke University's undergraduate environ-mental organization, Environmental Alliance, proposed energy conservation measures that were adopted by the Facilities Management Department and were later used to buy wind power. Student groups may also provide material resources by promoting an increase in student fees and using these fees to pay for the additional costs of wind power. This approach has been used by environmental groups at different universities such as the University of

Colorado at Boulder, the University of Pennsylvania, and Western Washington University.

Second, student environmental organizations may offer nonmaterial resources by distributing information about the benefits of renewable energy. Student groups such as Students for Renewable Energy at Western Washington University "learned everything they could about, not just wind energy, and efficiencies of that, but also the whole idea of green tag commodity markets."[36] They used this information to educate university administrators and to persuade the board of trustees that buying wind power certificates was feasible and had real benefits. According to one university official,

> They had it really well thought-out, well orchestrated, and well researched—and for that reason we were able to get [it] through. I think if it hadn't been so well researched we might have gotten derailed through the board of trustees. That's because the board of trustees, very rightly, had to be convinced that the premium required for wind energy was something that was real—in other words, that green tags were a real commodity and that we weren't simply throwing money down the drain.[37]

Similarly, in 2002 students from the Carolina Environmental Student Alliance at the University of North Carolina won a referendum proposing a student fee increase to fund renewable energy projects because they worked very hard to educate students on the sources of their energy.[38] And students from Green Arch at New York University made the decision to educate university administrators and avoid confrontational tactics with the administration because "the best way we could go forward was to present ourselves to the university as a resource. Basically, we did a lot of research, we gathered a lot of information, we reached out to administrators to whom the information would be valuable, and we gave them the information. We gave presentations all over campus, we met with them, we talked with them—it was really more of an educational sort of effort."[39]

The Campaign for College and University Presidents' Climate Change Commitment

In 2004, few students at Unity College, a very small and relatively young institution, were surprised that their clean energy campaign was a smashing success and that their college became the first in the state of Maine to use 100 percent renewable energy. After all, their college offers more environmental programs, and graduates more students with environmental majors, than any other college in the country. Many of its faculty members are environmentalists, and the president of the college is a self-declared environmentalist who drives a hybrid car. Moreover, a Sustainability Committee was formed as early as 1991, and students, faculty, and staff meet throughout the year and work to reduce the college's adverse impact on the environment.

The clean energy campaign at Unity College illustrates that college and university decisions to purchase wind power are influenced not only by the level of student mobilization but also by the presence of "environmental mediators"—faculty, staff, and administrators who, because they are sympathizers of the environmental movement and professional members of institutions of higher education, are in a good position to translate the appeals of student activists into greening the practices of colleges and universities.

Environmental mediators contribute to college and university decisions to buy wind power in three ways. First, top-level administrators may initiate the decision to purchase wind power certificates. This was the case at the University of Central Oklahoma, where the decision to buy wind power came from the vice president and the president of the university.[40] According to one administrator:

> As we were looking at what kind of university we want to be, we thought we would start on a more sustainable path and start to buy some of our energy from wind power—as a way to serve as an example for our country. Oklahoma is a very windy state, so if we could generate a little demand then there can be some more production and, eventually, we could bring the cost down. We also thought that it was a good example for students, being responsible corporate citizens in terms of a sustainable environment.... Once we got started we started hearing from a lot of student groups, we got a lot of encouragement from student groups and tremendously positive feedback, but initially it was not a student-driven initiative.[41]

Second, faculty members may inspire the emergence of a student campaign for clean energy by educating students about issues such as global climate change and challenging them to find ways to address those issues locally. Students at some schools, such as Western Washington University, started a clean energy campaign because they were encouraged by faculty to support renewable energy. According to one student organizer: "We were challenged by an environmental science teacher to do something with renewable energy and one of our group members learned about renewable energy certificates, and we decided we should go after that.... We connected quite early with sizeable environmental studies and environmental science classes, and we were able to do presentations in those classes about renewable energy certificates."[42] And the Step It Up National Day of Climate Action, the first "open source, web-based day of action dedicated to stopping climate change," was initiated by Bill McKibben—an environmentalist and educator at Middlebury College.

Third, faculty, staff, and administrators who act as environmental mediators may work closely with students who campaign for clean energy and decrease inertia to organizational change. For example, close collaboration between student activists and environmental mediators at the University of California at Santa Cruz resulted in the decision to offset 100 percent of the university's electricity use with RECs—the first such decision by a California university. As one university administrator observed, "The commitment to purchase renewable energy credits at UCSC, offsetting 100 percent of the campus electrical load, provides a leading example of how students, faculty, and administration can work collaboratively to reduce the environmental impacts of our institution. This effort has created a positive environment for future sustainability-related collaborative endeavors."[43]

Indeed, perhaps the most important contribution of environmental mediators is to reduce organizational resistance to change. This is clearly illustrated by, among others, the case of New York University, one of the universities that buys wind power for all its electricity consumption. According to one New York University student activist:

> One of the problems that I read and heard about from people at other universities was that it was difficult to convince people in the administration to move forward with these things. I was shocked, when the sustainability task force started meeting, by how incredibly receptive the university administration has been about this. They've gone out of their way, and I think it's obvious that they're not doing this simply because their students want them to do this, or because it's saving them money—because I don't think it's saving them money at this point. They're doing it because they really think it's the right thing to do, and it's been delightful to experience that over the past few months.[44]

Two environmental organizations, Second Nature, and the Association for the Advancement of Sustainability in Higher Education (AASHE), played an essential role in the emergence of a national campaign to mobilize top-level administrators and university presidents to address global climate change. Second Nature has been working since 1993 with administrators, faculty, staff, and students at colleges and universities across the country to make the principles of sustainability the foundation of all learning and practice. In 2001, it launched a program called the Education for Sustainability Western Network (EFS West) that focused on campuses in the western United States and Canada. EFS West became a membership organization in 2003, and in 2004 held the first North American Conference on Sustainability in Higher Education. Its leaders created AASHE to help coordinate and strengthen campus sustainability efforts at regional and national levels and to serve as the first North American professional association for those interested in advancing campus sustainability.[45]

At its annual meetings AASHE constantly promotes local actions against global climate change, and during the 2006 conference, it launched a program called the American College & University Presidents Climate Commitment (ACUPCC). Following this conference, 12 presidents agreed to become founding members of the Leadership Circle and sent a letter to nearly 400 of their peers inviting them to join the initiative. In a few months, over 150 presidents and chancellors representing a variety of colleges and universities had become charter signatories of the ACUPCC. Over 90 of them joined the Leadership Circle and agreed "to promote the initiative among their peers, serve as representatives to the press, and participate if possible in the public launch of the President's Climate Commitment in June. In late March, the expanded Leadership Circle sent a packet of information to their peers at over 3,500 institutions, asking them to sign the Commitment."[46]

Companies Switch to Renewable Energy

Between 1998 and 2008, the combined green-power purchases of large U.S. companies went from virtually zero to the equivalent amount of electricity needed to power many hundreds of thousands of average American homes each year. By January 2008, more than ninety companies and businesses had purchased RECs, mostly from wind power developers.[47] Of those, almost fifty companies were in the Fortune 500 (see table 4.2). The combined green-power purchases of those companies amounted in 2007 to more than 6.6 billion kilowatt-hours of green power annually, which is the amount of electricity needed to power more than 676,000 average American homes each year.[48]

Companies' decisions to purchase wind energy are influenced by the environmental movement in a number of ways. Environmental organizations and activists contribute to the emergence of environmental mediators, create a support network for climate change mediators, act as brokers who connect companies with renewable energy developers or utilities, certify the purchase of RECs, and pressure companies to address global climate change.

Similarly to colleges and universities, companies are likely to purchase energy from wind and other renewable sources if their employees are strongly committed to addressing global climate change. But in contrast to the case of colleges and universities, where the decision to purchase RECs has often resulted from a combination of mobilization from above (top-level administrators) and from below (students), in the case of companies the decision to purchase RECs has frequently come from mid-level company executives.

Most of the time, the idea for purchasing RECs originates among employees of departments of environmental affairs, not among "rank-and-file" employees or top executives. To implement this idea, environmental managers have to act as social movement organizers: they mobilize resources, use specific framing devices, and build support among key allies. One environmental manager from Mohawk Fine Papers, the largest premium-paper manufacturer in North America, emphasized the importance of framing the

Table 4.2. Fortune 500 Companies That Had Obtained Some of Their Electricity from Wind Power and Other Renewables by March 2008

Company	Source
3M	Wind, Biogas
Advanced Micro Devices	Wind, Biogas
Agilent Technologies	Wind
Apple Computers	Wind, Biogas
Applied Materials, Inc.	Wind, Biogas, Solar
Aramark Parks & Resorts	Wind, Biomass
Autoliv	Wind
Baxter International Inc.	Wind
Cisco Systems, Inc.	Wind, Biogas, Biomass, Solar
Citi	Wind
Dell Inc.	Wind, Biomass
DuPont Company	Wind, Biomass, Solar
FedEx Express	Solar
FedEx Kinko's	Various
General Electric	Wind
General Motors	Biogas
Hewlett-Packard	Various
IBM Corporation	Wind, Solar
Intel Corporation	Wind, Biomass, Geothermal, Solar
John Deere Co.	Wind
Johnson & Johnson	Wind, Biomass, Hydro, Solar
Kohl's Department Stores	Various
Liz Claiborne, Inc.	Various
Lockheed Martin	Wind
Lowe's	Biogas, Solar
Macy's, Inc.	Solar
Monsanto	Wind
Nike, Inc.	Various
Office Depot	Wind, Biomass, Solar
Oracle Corporation	Wind, Biogas
PepsiCo	Various
Pitney Bowes	Wind, Biomass
Raytheon	Wind
Roche	Wind, Solar
Safeway Inc.	Wind
Sprint Nextel	Wind
Staples	Wind, Biomass, Solar
Starbucks	Wind
State Farm	Wind, Biogas
State Street Corporation	Wind
The Coca-Cola Company	Wind
The Estée Lauder Companies	Wind
The Pepsi Bottling Group, Inc.	Various
Time Warner Cable	Wind, Biogas

(*continued*)

Table 4.2.   (*Continued*)

Company	Source
United Parcel Service	Biomass, Solar
Wells Fargo & Company	Wind
Whole Foods Market	Wind, Biogas, Solar
Yahoo! Inc.	Wind, Solar

Source: U.S. Environmental Protection Agency, Green Power Partnership.

purchasing of RECs as a "goodwill investment," and not as a short-term expense, in order to overcome organizational resistance. He also emphasized the importance of winning support from top-level managers: "Initially this decision was driven by people in the Environmental Affairs Department. It was met with some degree of skepticism, but once we gained the support of a few key people in the marketing area and the support of the CEO, we got it off the ground. Goodwill is an invaluable commodity in business; once this project was recognized for its goodwill potential, any previous resistance faded away."[49]

The importance of having allies in other departments as well as among top executives was also emphasized by an IBM employee. He emphasized that, while having the company commit to addressing climate change is important, environmental managers have to build local coalitions that support practices such as green-power purchases in order to move from formal recognition of the problem to actual steps. As he put it:

IBM recognizes that climate change is an important environmental issue and a challenge that needs attention. We try to show that it is important to begin taking action now, recognizing that the whole palette of solutions is going to change and develop over time, but that it is logical to begin to take actions today for developing that palette and to make some inroads on reducing emissions. The primary focus on that comes from the corporate environmental staff, and we set the policy and advocate for specific activities within the group and the company. On energy issues we work hand in hand with the real estate and site operations groups that do the majority of the sourcing for energy around the globe. The primary advocacy comes out of the environmental group, but there has been strong support in the real estate group to take incremental steps and look for places where it makes logical business sense for us to make these sourcing decisions. There are also country-level executives who are engaged in the process when sourcing decisions are made, who have chosen to source out some percentage of renewable energy for business reasons within that geography. So, you have got essentially a group of advocates that make it happen.[50]

Other interviewees also highlight how important it is for environmental managers to receive support from senior executives. While CEOs and other top-level managers are rarely the architects of environmental practices such as green-power purchases, their support is essential in order to implement the practices and gradually expand the percentage of green power purchased by companies. According to an energy specialist who has worked with firms that purchase RECs, successful implementation depends on a number of things:

> Firstly, you need senior management support. It takes a senior executive like a CEO to say, "We're going to take climate change seriously" or "I'm going to support my energy team to diversify into renewables, away from just natural gas or coal." So, senior management support is very important. Secondly, it depends where the environment is on the company agenda—which is similar to the first one. As companies roll out their sustainability programs, the ones that have a strong commitment to addressing climate change are most likely to look into renewables.[51]

In addition to framing the purchasing of RECs as a long-term "goodwill investment" and to maneuvering through corporate channels to find influential allies, climate change mediators also have to mobilize company resources. In most cases, the decision to buy electricity from wind and other renewable sources follows the adoption of practices aimed at increasing energy efficiency. Because increasing energy efficiency usually results in reductions in electricity expenses, climate change mediators have an easier task in getting support for practices that require a premium for green power. Consider how one environmental manager from Johnson & Johnson describes long-term efforts to reduce greenhouse gases, which culminated with the purchase of RECs:

> Back in 1999, we recognized that climate change is an important issue because of the impact it could have on human health. Back then we set a goal and, although the U.S. wasn't part of Kyoto, we announced a reduction of emissions by 11 percent in all our facilities worldwide by 2010 compared to 1990. Once we made that decision to set that goal and to reduce our emissions, it was a matter of creating a strategy as to how to do that—and back in 1999 we really didn't know exactly how we would accomplish that. The first thing we did is really focus on efficiency; so we have a set of best practices and we want all of our facilities to meet those best practices. That is just things like most efficient lighting, efficient equipment, and so on. So that was fairly straightforward and that's where most of our benefit comes from. Then the other piece was to go out and purchase renewable energy wherever we can. In 2003, we actually made this a policy of the company, which was made public and our board of directors approved it. We have been implementing that general strategy since then.[52]

Some companies also purchase green power because their employees like to work for companies that care not only about their shareholders but also about the environment. Indeed, feeling proud of working for a company that listens to its employees and is doing "the right thing" was a recurrent theme in many interviews. Consider the following statement from an environmental manager working for Whole Foods:

> We operate as a consensus-driven organization. The decision [to purchase RECs] went through the channels of a national task force meeting, all the regional presidents signed off on it. I guess to me, as a team member working there, it just makes me more proud of what I'm doing and feeling like I'm committed to a company that is doing the right thing. A lot of my work is public facing, I go to events and I'm always happy to talk about this purchase. It helps customers and people in the community understand what green power is and why it's important to purchase it.[53]

The fact that the decision to purchase wind power was a moral decision, not a business decision, is emphasized by another Whole Foods employee. He emphasizes that it was possible to implement his plan for a major wind power purchase because his company is committed to environmental stewardship and has a democratic structure, which allows for input from all employees:

> The Green Mission taskforce was me and then anyone else who wanted to volunteer. And that was the case for a couple of years because of the way the company is structured—with an empowerment culture that comes from the bottom level rather than the top. We had previously installed some solar roofs, but that wasn't enough. I created a committee that looked at the possibility of doing more, and there were lots of opportunities. The only one that at the time presented the solution for 100 percent offset was wind. That's where we went. The reason is because it is part of our core values, rather than a response.... The decision, as far as a business decision, was made on the basis of "It's the right thing to do" rather than "What would the return be?" You can't calculate the returns when you're doing something like this; it's not the same as projecting sales and there was no way that you could project that sales were going to be greater and that there would be an actual, real, tangible increase in revenues. It's something that we, as a company that prides itself on the core values of environmental stewardship, had to do.[54]

But corporate decisions to purchase renewable energy are influenced not only by the presence of environmental managers or internal wind power champions. They are also influenced by a small number of environmental groups. One such organization is the World Resources Institute (WRI), which was formed in 1982 through a grant from the John D. and Catherine T. MacArthur Foundation.[55] As the climate change issue became the dominant environmental problem on the international agenda during the 1990s, the WRI launched an action agenda in 1998 under the title Safe Climate, Sound Business.

Through this collaborative effort with General Motors, Monsanto, and British Petroleum, the WRI intended to show that addressing global climate change and promoting economic growth are not incompatible policy goals. The action agenda stated that "although addressing climate change will be a challenge, we believe that there should be no inherent conflicts between economic development and a healthy environment." Furthermore, the agenda highlighted three conclusions: (1) "Climate change is a cause for concern, and precautionary action is justified now," (2) "Business can contribute to climate protection efforts in substantial, positive ways by helping to develop sound climate policies, by providing the research and technologies needed to address the challenge, and by taking actions to reduce and offset their own emissions," and (3) "Flexible and market-oriented climate policies that implement national commitments can address the long-term need to stabilize the concentration of greenhouse gases."[56]

The WRI's Safe Climate, Sound Business action agenda was followed in 2000 by a program called the Green Power Market Development Group, a commercial and industrial partnership dedicated to building voluntary markets for green power. This program started from WRI energy experts' observation that

> more than 80% of man-made U.S. greenhouse gas (GHG) emissions arise from burning fossil fuels such as coal, oil, and natural gas. Switching to renewable energy or "green power" would significantly reduce these energy-related GHG emissions and therefore is an attractive strategy for combating climate change. In particular, green power will need to dramatically penetrate the commercial and industrial sector since corporations currently account for over 50% of U.S. energy consumption.[57]

Consequently, the program had two goals: a general goal "to enable corporate buyers to diversify their energy portfolios with green power and reduce their impact on climate change," and a specific goal "to develop 1,000 megawatts of new, cost-competitive green power by 2010 in the U.S."

The WRI was important for the growth of the nonresidential market for renewable energy for a number of reasons. First, the WRI raised awareness of global climate change and other environmental problems among employees of large corporations. Indeed, environmental organizations such as the Nature Conservancy, the World Business Council for Sustainable Development, the Coalition for Environmentally Responsible Economies, and the WRI have attempted to bring about a "sustainability revolution" not by pressuring companies from outside, using boycotts or protests, but by creating change from inside.

The efforts of the WRI and other organizations are partly responsible for a major change in corporate culture—from an emphasis on shareholder profit to one on social responsibility toward various stakeholders. As one author argues, "not since the Industrial Revolution of the mid-18th and mid-19th centuries has such a profound transformation with worldwide impact emerged onto the world stage. Like its industrial counterpart, the Sustainability Revolution is creating a pervasive and permanent shift in consciousness and worldview affecting all facets of society."[58] A 2002 survey of managers at various U.S. corporations found that almost 60 percent of the companies had a structured program for engaging with various stakeholders on a regular basis, that 75 percent of the companies had a relationship with one or more NGOs, and that the largest number of these relationships—one in five—were with environmental groups.[59] Studies of mass media also show that companies have a growing interest in sustainability; for example, the number of "green" stories in newspaper business sections increased from less than 40 in 2000 to more than 180 in 2007.[60] And in May 2005, the CEO of General Electric announced that his corporation would be staking its future on the ability to "define the cutting edge in cleaner power and environmental technology"—a statement considered by some journalists as "the most dramatic example yet of a green revolution that is quietly transforming global business."[61]

Second, the WRI was important for the growth of the nonresidential market for renewable energy because it created a support network for corporate climate change mediators. Following the Safe Climate, Sound Business action agenda, the WRI began actively recruiting a number of companies that had expressed interest in limiting emissions of greenhouse gases. One WRI energy expert describes the formation of the Green Power Market Development Group this way:

> One of the points on the action agenda was to get companies to increase use of renewable energy and one of the companies said, "Well, we don't know how to do that; why don't you (WRI) do something to help the corporate sector out?" We said, "OK, let's actually form a renewable energy buyers group and you can be a member," and they said, "Sure." This company recruited a couple of companies they knew; we then went out and recruited a few more of companies. We started off in 2000 with ten firms but later we recruited more members and we gradually expanded.[62]

The Green Power Market Development Group makes it easier for climate change mediators to overcome organizational inertia toward purchasing wind power and other types of renewable energy. Environmental mediators working for companies that join the group have easy access to information about renewable energy's benefits. For example, in 2005 the WRI distributed a document titled "The Business Case for Using Renewable Energy," which highlighted that by switching to renewable energy, companies can reduce emissions of greenhouse gases and other airborne pollutants that pose regulatory risks, stabilize corporate energy prices, reduce operating losses caused by power outages, and strengthen company relationships with various stakeholders. These environmental mediators also have more clout with company decision makers and can persuade them that purchasing renewable energy is necessary in order to keep up with other leading companies. As one IBM employee notes,

> One benefit of joining the Group was the information exchange about different kinds of projects that were being done at different companies—that gave us some idea of the full range of opportunities. Another one was that it provided us with support when we advocated internally for renewable energy—being able to say that other leading companies are also taking these kinds of actions. The ability to say, "We are not the only ones out there doing this" provided us with an important justification.[63]

Third, the WRI promoted nonresidential renewable energy use by acting as a broker that connects companies with renewable energy developers or utilities and negotiates the best deals. This is how one WRI energy expert describes his organization's role:

> What we do with the companies is we start off with bringing them up to speed on the technologies, how do they work, what are the economics, how do you do a project, what are the barriers, who are the developers, who are the suppliers, etc. There is that learning curve piece and we work with the companies to develop a business case, teach them why they should do this. Then, for those who actually pursue projects, we act as the middle person: we'll match buyer and seller, we'll introduce companies to their suppliers of green power, of RECs or of on-site systems. We meet three to four times a year, and we'll bring in developers and suppliers to a meeting to introduce one to the other so the companies can ask the developers what's the real economics of doing an on-site solar PV [photovoltaic] system or what states are the best to do a solar PV system in, what states have the best incentives for wind power, etc.[64]

Companies that belong to the Green Power Market Development Group benefit from the WRI's experience in negotiating group purchases of RECs. According to one Johnson & Johnson employee, one of the main benefits of joining the group is that "the WRI does some work to accumulate all our interests in making purchases and doing some marketing work to drive the price based on the volume the group can bring."[65] In conclusion, the WRI has played an essential role in the growth of the corporate wind power market in the United States. Indeed, by August 2007, the WRI's Green Power Market Development Group had reached 738 MW of new renewable energy generation, mostly from wind power facilities—only 262 MW away from achieving its goal of developing 1 GW of new green power by 2010 in the United States.[66]

Other regional and national environmental organizations have also had important functions for the emergence of a voluntary green-power market. Clean Air-Cool Planet (CA-CP), for example, has partnered with companies throughout the Northeast to help reduce their emissions of greenhouse gases. In 2002, CA-CP entered a partnership with Timberland "to undertake an inventory of its greenhouse gas emissions, establish a reduction target, and help educate its suppliers, vendors, employees, and ultimately customers on the economic benefits of taking action to address climate change."[67] Because of this partnership, Timberland has implemented a variety of energy-efficiency programs and has purchased a significant number of RECs. Similarly, the World Wildlife Fund has started a Climate Savers initiative to get companies to establish ambitious targets to voluntarily reduce their greenhouse gas emissions. Johnson & Johnson, one of the early partners in this program, has purchased a significant number of RECs as part of its commitment to reduce its greenhouse gas emissions from all facilities worldwide to 7 percent below 1990 levels by 2010.[68]

Environmental organizations have also stimulated the growth of the corporate green-power market by certifying the purchase of RECs. The Center for Resource Solutions (CRS), for example, was founded in 1997 with the mission to "help lead the industry in the design and implementation of programs to increase the demand and use of renewable energy around the world."[69] As many utilities began selling renewable energy in California and other states, the CRS realized that no standards existed to ensure responsible renewable energy production and sales. Therefore, the CRS created the Green-e program with its fourfold mission:

> [To] bolster customer confidence in the reliability of retail electricity products reflecting renewable energy generation. Expand the retail market for electricity products incorporating renewable energy, including expanding the demand for new renewable energy generation. Provide customers clear information about retail clean electricity products to enable them to make informed purchasing decisions. Encourage the deployment of electricity products that minimize air pollution and reduce greenhouse gas emissions.[70]

Green-e has a reputation for high quality, and many U.S. companies that purchase RECs use it to verify that the energy they buy comes from new renewable resources and to quantify reductions in greenhouse gas emissions. For example, the Green-e program allowed Wells Fargo to claim that its purchase of renewable energy credits stimulated the development of new wind energy projects in the United States and had a major impact on the environment. Using support from the CRS, Wells Fargo calculated that its purchase of wind power prevented the emission of "380,000 tons of carbon dioxide each year, the equivalent of reducing the CO_2 emissions of 75,000 cars annually or by reducing the equivalent CO_2 emissions associated with 40,000,000 gallons of gasoline each year."[71]

Still other environmental organizations have contributed to the growth of the corporate green-power market through direct actions such as protests and demonstrations. Rainforest Action Network (RAN), for example, had an important role in Wells Fargo's decision to purchase 40 percent of its electricity from Green-e certified wind energy—the largest corporate purchase of renewable energy in the United States by 2007. Although the president of Wells Fargo announced that the purchase "demonstrates our company's commitment to both environmental stewardship and environmental leadership, and it reflects the desire of our team members to do what's right for our customers, our communities, and our company," in reality Wells Fargo's desire to do "what's right" was partly influenced by RAN's actions.

Rainforest Action Network activists accused Wells Fargo of funding projects that increased global warming, and used street theater, leaflets, and banners to spread the word that Wells Fargo was the largest U.S. bank operating without a comprehensive environmental policy. Rainforest Action Network's strategy consisted mostly of pressuring Wells Fargo to change its practices by embarrassing its investors. According to one RAN activist quoted in the media: "Wells Fargo may not care that investing in mountaintop coal removal in Appalachia destroys communities and the environment, but we thought its investors would. Investors are in a perfect position to tell Wells Fargo to stop investing in destruction and start investing in the future."[72] Partly because of Rainforest Action Network actions, in March 2006 the company formed a new Environmental Advisory Board, published a Corporate Citizen Report in which it detailed its efforts to reduce greenhouse gas emissions, and announced that it would purchase 40 percent of its electricity from wind power.

Finally, other environmental groups have specialized in pressuring companies through shareholder activism to become greener. For example, CERES, a national network of investors, public interest groups, and major environmental organizations such as the Sierra Club, the Union of Concerned Scientists, Environmental Defense, the World Wildlife Fund, and Friends of the Earth has been working with companies and investors to address sustainability challenges such as global climate change. Rather than organizing protests, CERES has pressured companies mostly by mobilizing shareholders to adopt resolutions for greening corporate practices. In 2007, CERES published a "climate change blacklist" in which it accused ten companies—including the oil giant ExxonMobil, the financial services group Wells Fargo, and the utility TXU—of not doing enough to respond to global warming. As the president of CERES stated: "Many U.S. companies are confronting the risks and opportunities from climate change, but others are not responding adequately—and they may be compromising their long-term competitiveness and shareholder value as a result. We want all companies to understand the business impacts of climate change—and plan for it accordingly. It's what any corporate director would expect of their CEO."[73] Apparently, the CERES strategy of mobilizing shareholder activists to pressure company executives to address global climate change has played a nontrivial role in some companies' decisions to purchase major amounts of RECs.[74]

Conclusion

This chapter has shown that, although American environmental groups have had little success in influencing federal energy policies, they contributed to a significant increase in local demand for renewable energy. It has demonstrated that environmental organizations played key roles in the decisions of colleges, universities, and corporations to purchase RECs from wind and other renewable energy sources. The combined effect of these green-power purchases is far from negligible: over one decade, they account for over 3.1 GW of new renewable energy capacity additions, mostly from wind power.

The chapter has also showed how environmental groups shape organizations' decisions to purchase green power. Many environmental groups offer crucial mobilizing resources for green-power champions. Others act as brokers who connect organizations with renewable energy developers or utilities, as certification agents who verify the purchase of RECs, or as organizers of protests, boycotts, or shareholder activism. The chapter's analysis has demonstrated that, while environmental groups and activists can sometimes pressure organizations to change "from the outside" through protests, boycotts, and lawsuits, their most significant impact is to create change "from the inside." The environmental movement's main impact is to transform organizational members into environmental mediators. In the case of colleges and universities, national and local environmental groups have pushed for green-power purchases both bottom-up, by organizing student campaigns for clean energy, and top-down, by coordinating a network of college and university presidents who are committed to addressing climate change. In the case of companies, environmental groups have pushed for green-power purchases mostly from the center, by offering resources to mid-level employees and environmental managers.

As a final note, although this chapter has focused only on the nonresidential green-power market, it is important to point out that the U.S. environmental movement has also influenced the growth of the wind energy industry by stimulating the growth of the residential green-power market. Virtually all major environmental organizations encourage their members to take individual steps to reduce emissions of greenhouse gases and offer tips for "things you can do" to address global warming. The organization Stop Global Warming, for example, promotes not only actions that result in protecting the environment and saving money over the long term (such as using compact fluorescent bulbs, adjusting the thermostat, or insulating the water heater), but also actions that require paying a premium (such as buying wind certificates and green tags).[75] Thus, while the chapter has focused only on the effect of the environmental movement on universities and corporations, it is important to observe that the movement has also contributed to the growing demand for wind energy by changing personal lifestyles and behaviors.

This chapter has focused only on the case of the United States, yet residential and nonresidential green-power markets also exist in countries such as the Netherlands, the United Kingdom, and Germany. In the Netherlands, green-marketing programs became so popular at the end of the twentieth century that utilities had difficulties in keeping up with demand (Redlinger, Andersen, and Morthorst 2002, 179). In the United Kingdom, utilities such as Ecotricity have been selling wind power to individuals and businesses since the late 1990s.[76] And in Germany, Greenpeace has been selling green power to over eighty thousand of its members through Greenpeace Energy.[77] The next chapter will examine yet another pathway of influence of the environmental movement: the restructuring of the energy sector.

5

Going with the Wind

The Environmentalist Transformation of the Electricity Sector

Our customers and local communities expect us to protect the environment, and as the world becomes increasingly concerned with global climate change, environmental leadership has grown in importance for our shareholders, employees and the future of our company. Environmental leadership means that as we provide energy services to our customers, we will pursue clean energy innovation, transforming how energy is provided.
—Xcel Energy, http://www.xcelenergy.com/SiteCollectionDocuments/ docs/2007TBLFull.pdf#page=4 (accessed December 2008)

The Greening of the Electricity Sector

Xcel Energy, the fourth-largest natural gas and electricity company in the United States, started investing in wind power in 1997.[1] In early 1998, Xcel already operated thirteen wind turbines and had signed up more than four thousand of its customers to purchase 9.6 MW of wind power through its Windsource program. By 2001, Xcel operated wind farms that could produce up to 60 MW of wind power and had more than seventeen thousand wind power customers. By 2008, Xcel operated 2.7 GW of wind power and had a program that was the number one voluntary green-energy program in the country, with more than seventy thousand customers participating.[2]

Although Xcel was one of the first major electric utilities in the world to develop or operate wind farms, more and more utilities are doing this nowadays. In many countries, the electricity sector has undergone two significant transformations over the last two decades.[3] First, wind turbine manufacturing has recently become big business in countries such as Denmark, Germany, and the United States. Small, traditional wind turbine manufacturers from Denmark, such as Vestas, have become industrial heavyweights. At the same time, a growing number of giant energy companies in Germany and the United States, such as Siemens and General Electric, have entered the market for wind turbine manufacturing.

Second, large electric utilities have increasingly become developers or owners of wind power projects. Some electric utilities, such as Spain's Iberdrola or the United States' FPL Energy, have developed a large renewable energy portfolio and have become the world's leading wind-farm operators. Other utilities, such as Great Britain's Npower, Germany's E.ON, and the United States' Duke Energy, have only recently begun to build or buy wind projects. New companies that specialize in wind power projects have also emerged as important players: examples include the United Kingdom's Ecotricity and the United States' Community Energy.

What accounts for these important transformations of the electricity sector in different countries? Improved economics, obviously, is important: wind turbines have become much larger and more reliable, while the capital costs of building a wind turbine have come down. Compared to solar power, wind energy is a cheaper and significantly less risky form of investment.[4] Even compared to fossil fuels, wind power has recently come close to being competitive: under special circumstances, wind power can be cheaper than natural gas or oil.[5] But this account does not fully explain why power plant manufacturers in particular regions or countries dominate the global market for wind turbines, nor does it explain why only certain electric utilities invest in wind power.

This chapter shows that the most remarkable transformations of the electricity sector happen when environmental activists and sympathizers are able to exert influence on energy companies and professional societies, critique the traditional logic of energy production, and offer a solution—hinging on an environmentalist logic—to the electricity sector's problems. The environmental movement stimulates industrial activity when environmentalist norms and cultural frameworks shape wind energy entrepreneurs' perception of social opportunities and motivation to take risks to exploit these opportunities. Environmental movement activists and sympathizers may contribute to wind turbine manufacturing by becoming entrepreneurs, innovators, advocates, or champions.[6] Furthermore, environmental groups and activists may pressure utility companies to invest in renewable energy by using tactics such as protests, lawsuits, and lobbying for stricter regulation, may form new companies that specialize in wind-farm development and operation, and may aid developers overcome local opposition to wind farms. The next sections focus on the role played by environmental activism in wind turbine manufacturing and wind-farm development.

Environmental Activism and Commercial Wind Turbine Manufacturing

During the 1980s, commercial wind turbines were manufactured by small Danish companies that were energy sector newcomers. Four of the top five wind turbine manufacturers at the end of the 1980s were relatively small Danish companies: Vestas, which produced agricultural vehicles and hydraulic cranes; Nordtank, which produced container transportation vehicles; Danregn (later know as Bonus), which produced irrigation equipment; and Micon, a new company founded by a former Nordtank employee. Starting in the early 1990s, companies building commercial wind turbines grew through mergers and acquisitions. In 1997, Nordtank and Micon merged and formed NEG Micon; in 2004, Vestas and NEG Micon merged under the name Vestas, becoming the world's largest wind turbine manufacturer.[7]

German companies also emerged as major wind turbine manufacturers during the 1990s. By 1997, the German companies Enercon, Tacke, and Nordex were three of the top ten wind turbine manufacturers worldwide.[8] By the beginning of the twenty-first century, companies that had traditionally manufactured fossil-fuel and nuclear power plant equipment had also begun manufacturing wind turbines. This was the case with General Electric, which began manufacturing wind turbines in 2002 (after acquiring Enron Wind), and Siemens, which began manufacturing wind turbines in 2004 (after acquiring Bonus). While major suppliers of wind turbines also emerged in countries such as Spain, India, and China, the market for wind turbines continued to be dominated by Danish and German companies; for example, at the beginning of 2008, Vestas, Enercon, Siemens, and Nordex supplied almost 50 percent of the wind turbines marketed worldwide.[9]

The question, why were the Danes best? has already been asked in a number of studies that have focused on differences in national technological styles and corporate approaches to technology.[10] This section demonstrates that it is not possible to fully answer this question without examining the role of the Danish environmental movement. It also shows that the environmental movement has contributed to the emergence and consolidation of wind turbine companies in countries such as Germany and the United States.

Environmental Activism and Wind Turbine Manufacturing in Denmark

One of the most important wind energy entrepreneurs and innovators in Denmark was Erik Grove-Nielsen, who became involved in the grassroots campaign against nuclear energy as a student. In 1971, Grove-Nielsen became a member of a student environmental organization at the Technical University in Copenhagen. Being technically skilled, he became interested in developing sustainable solutions to energy problems and put up a solar hot water collector on his house. He interrupted his studies and decided to do something to "change the world" when the energy crisis of 1973 hit Denmark and the rest of the world; as he recollects:

> I was always involved in environmental thinking. When living and studying civil engineering in Copenhagen, I was working with solar; in '71 I put up a solar collector on the top of our house there. In 1973, I stopped studying there because I wanted to do something to change the world, so I went back to the western part of Denmark, where I was going to start up the production of aluminum solar collectors, but I didn't have luck with that. Then in '74, when electricity companies wanted to put up nuclear power, I was working half-time in a daytime job and the other half of the time volunteering in OOA [Organization for Information about Nuclear Power], and there was a local group in Viborg, close to where we lived. Then I joined this grassroots group of people, and we were discussing what we could do to avoid nuclear power in Denmark.[11]

While he was volunteering for OOA, Grove-Nielsen became interested in developing alternatives to nuclear energy and became a founding member of the Organization for Renewable Energy (OVE). He learned about wind energy through regular interactions with other OVE members. Indeed, OVE had an essential role in getting young, idealistic people like Grove-Nielsen involved in wind turbine manufacturing. The Organization for Renewable Energy connected people who were interested in renewable energy and created a nationwide network of wind energy pioneers by publishing a series of books with their names, contacts, and other useful information. According to Grove-Nielsen, OVE

> collected the names and addresses and phone numbers—there was, of course, no Internet at that time... One guy was driving from town to town asking, "Do you know about people working with solar, or wind, or bio?" He collected addresses and went on to the next place. He made a book with addresses of these "grassroots engineers" or what-ever you call them, people working with renewables. There were also a few very small companies or self-builders who worked with different technologies. I think the first book was made in '76, and then another one came one year after. There were two or three very important books, so we could find each other. And then there was a book called the "Solar and Wind Handbook." That book would be on every person's shelf—for those working with solar or wind.[12]

In addition to publishing books on renewable energy, OVE organized meetings at which environmentalists like Grove-Nielsen could learn about each other's work and make contacts with energy professionals from research institutes and universities, as well as with companies interested in wind turbine manufacturing. These meetings were organized in the tradition of the Danish folk high schools, in which people from various backgrounds interacted and exchanged opinions freely—an essential condition for the dissemination of knowledge about wind and other renewable energy technologies. Here is how Grove-Nielsen describes the first meetings:

> In '76 we started having meetings two or three times a year, and then everyone who wanted to work with the wind energy industry would come there to discuss what happened, and also guys from the Risø institute and guys from the university would be there together with people living on farms because most of this happened in western Jutland, in the farm belt and not in the big cities. So OVE had these meetings and when new companies wanted to join, they would meet the grassroots people at those meetings.[13]

During the OVE meetings, Grove-Nielsen learned that one of the most common technical problems faced by small-scale wind turbine manufacturers was that of blade reliability. Sensing an entrepreneurial opportunity, he decided to form his own blade manufacturing company. Because he knew many other wind energy pioneers through OVE, he was able to buy the molds for the first fiberglass blades made in Denmark, sold to him by the people who built the Tvind turbine. He recollects:

> We saw from these OVE meetings there was a problem with the blades because people made wood blades, they would make welded steel blades but many of them came flying off the turbines. I was familiar with some wind aerodynamics so I thought, "Why not make wind turbine blades that are effective?" But I didn't know anything about glass fiber at that time. The first thing was building these small 1.7 meter long wind turbine blades. Then, in the summer of '77, I decided to start a company. What happened then was there was a turbine in the southern part of Denmark that was made by a small company, and they had a blade from a small Tvind turbine. Before Tvind made the big blades, they made small blades to start with. But those blades that the company in southern Jutland had made had a big flaw in the root, so just after a half a day or so they came down to the ground.[14]

Grove-Nielsen purchased the mold that was manufactured at Tvind for a small price (approximately US$400) and founded Økær Vind Energi in August 1977. The company was a two-person operation on a shoestring budget; Nielsen built up a fiberglass production facility in his old farm building. He obtained a bank loan of approximately US$10,000 and began manufacturing blades that he sold to Svend Adolfsen's company (which later became Kuriant), the first Danish company to manufacture a grid-connected wind turbine in serial production during the 1970s.[15] However, Grove-Nielsen could not sell blades to other wind turbine manufacturers even after advertising in a major Danish newspaper and driving around the country in his Volkswagen van "with a blade pointing out of the rear hatch." He soon realized that he had two problems: first, his blades were too fast-running and made a lot of noise; second, none of the blacksmiths who were building wind turbines were prepared to start large-scale production. His temporary salvation came when he was contacted by Preben Maegaard, who led a small local association for sustainable energy called the Northwest Jutland Institute for Renewable Energy. Preben gathered four wind energy enthusiasts who wanted to build their own turbines and asked for a different blade design—he wanted blades that were five meters long and had a tip speed of around forty meters per second.

Between 1978 and 1980, Grove-Nielsen sold the new five-meter blades to self-builders all around Denmark, as well as in Sweden and Germany. At the time, no companies involved in wind turbine manufacturing existed in the world. By 1980, however, approximately twenty start-up companies "popped up like mushrooms" in Denmark, and many were formed by environmental activists like Grove-Nielsen.[16] During this period, Økær Vind Energi experienced numerous technological and financial challenges. Because the first blades were not equipped with tip-brakes, turbines would spin out of control and self-destruct when the wind was too strong and the mechanical brakes failed. After two such incidents, Grove-Nielsen stopped production and devised a new system for air braking with springs. But his company soon ran into financial difficulties, and as Grove-Nielsen says, only his personal connection with OVE saved his company:

> As my production was halted, the invoices kept coming but no income was created. Furthermore, the motor of our Volkswagen transporter broke down. The bank did not want to loan us money for a spare motor. In despair, I contacted Lars Albertsen, of the grassroots organization OVE.... One week later, a check of 50,000 Danish kroner (US$ 8.000) came in our letterbox. This money saved our company from bankruptcy. We did not lose our home, the new safety system could be developed, and we could have our VW running again.[17]

Grove-Nielsen's company survived during the difficult times at the end of the 1970s because he was determined to "do the right thing" for the environment and future generations. He recollects how, while driving across the country in search of customers, he saw the Swedish nuclear power station at Barsebäck and stopped to take a picture with the nuclear plant in the background and his Volkswagen van in the foreground, with wind turbine blades pointing over the nuclear plant. He saw this as a symbol of the victory of the environmentally friendly wind energy industry over nuclear power. As he said about the Barsebäck power plant, "Decommissioning of the plant will begin in 2020 and is planned to be finished by 2027. It is good to know that decommissioning of wind turbines takes place in 1–2 days (for land based turbines, and probably a few weeks for offshore turbines). Every ton of material can be recycled into new products. No waste is left over as a burden to our children."[18]

After the difficult start-up period, Økær Vind Energi started to grow. At the end of 1979, the company employed four people; two years later it employed thirteen people. In 1980, the company started to sell the five-meter blades to three wind turbine manufacturing companies that would soon become market leaders: Vestas, Nordtank, and Bonus. The same year, Økær Vind Energi started producing 7.5 meter blades for the Vestas 55 kW turbine. As Nordtank and Bonus also started to build 55 kW wind turbines for the California market, Økær Vind Energi became the main supplier of blades for the nascent global wind energy industry. In 1981, the company also made the first blades for Aloys Wobben, the founder of Enercon—which later became the largest wind turbine manufacturer in Germany.[19] At the end of 1980, Grove-Nielsen bought an industrial building in a nearby village and equipped it for fiberglass blade production. Because the company had grown, it began employing not only environmental activists involved in the grass-roots movement for renewable energy but also union members who required better working conditions.

But, just as the company was expanding and growing beyond the environmental activist community, it suffered again from a string of technical and financial problems. Two of the 7.5 meter blades sold to Vestas failed, and Økær employees had to revise the design in cooperation with technicians from Vestas and the Risø National Laboratory. The production of the revised 7.5 meter blade was resumed in early 1981, but the company had lost a lot of money because it had to replace all the blades on the 55 kW wind machines. For Grove-Nielsen, the situation was simultaneously promising and frustrating. On the one hand, the California "wind rush" created a rapidly growing demand for blades from companies that used to manufacture agricultural equipment but had switched to wind turbine manufacturing—Vestas, Nordtank, and Bonus. On the other hand, the redesign and replacement of the old 7.5 meter blades had had a big impact on his company's finances, and he couldn't expand capacity. The financial situation was so bad that when Grove-Nielsen could not pay the electric bill for his company and house, the electricity supply was cut and his family had to use a bucket on a rope to get water from a well. Therefore, he negotiated a license agreement with the owner of a fiberglass-boat manufacturing company in May 1981; his company stopped production and went bankrupt shortly thereafter.

From 1981 onward, the blades designed by Grove-Nielsen became known as AeroStar blades and were manufactured by Coronet—a company that was later reorganized and renamed, first as Alternegy, then as Danish Commercial Energy Research. Former Økær Vind Energi employees trained Coronet workers who were skilled in fiberglass use but not in blade manufacturing. While companies such as Vestas started designing and manufacturing their own blades, they also kept purchasing AeroStar blades until the mid-1980s. In December 1986, the company Danish Commercial Energy Research, which produced the AeroStar blades, went bankrupt. As the California market collapsed in 1986, most of the Danish wind power companies—including Vestas and Nordtank, but with the exception of Bonus—went bankrupt.

During this difficult period, Grove-Nielsen established a blade-fatigue test facility. Although he was not involved in manufacturing wind turbine blades anymore, he remained active in the industry by performing fatigue tests for large blades. In 1991, the Risø institute—the largest research institute in Denmark—acquired the technical equipment from Nielsen's blade-fatigue test facility, and he became employed as the head of the Risø blade-test facility. He left this position in 2000 to found a new company, ReFiber, which introduced innovative techniques for recycling fiberglass waste from wind turbines. Because he was not able to attract investment for a large recycling facility, in 2007 Nielsen started working as an external consultant for the blade technology unit of Siemens Wind Power.

Another very important wind energy activist-entrepreneur was Henrik Stiesdal. Although not involved in OOA (he was too young at the time), Stiesdal was inspired by the grassroots activists of the 1970s and, in particular, by the construction of the Tvind turbine. He remembers being very impressed by the people who were building a very large turbine near his hometown:

> I went traveling after I had left high school, and when I came back home at Christmas in '76 my father said, "While you have been away, a strange thing has happened in the neighborhood: they've started building this wind turbine. Let's go up and have a look." So, there were all these young people who were involved and who were working basically on their own with home-built equipment and building a home-designed huge machine. And for me it was a very big inspiration; I thought, "Well, we can do it. If we want it enough, we can do it. Other people are not doing this, but we want it to happen, so we can make it happen." That was a very strong piece of inspiration.[20]

Because both he and his father were technologically skilled, they helped the antinuclear activists with the construction of a smaller wind turbine:

> During 1977 we got more involved with them, not so much with the big turbine, but my father and I built some control equipment for a smaller turbine that they also made because there were things that, no matter how much they wanted to do, they could not do. My father was a physics teacher, so he knew something about controls and so on, so he and I made a control system for their turbine. That way we got to be somewhat more involved with them for a period.... After a while, our interest in working with them declined because we felt they were not, so to speak, of a democratic outlook on life. But in the early years, when that had not been realized, it was hugely interesting and very inspiring.[21]

Stiesdal remembers that he learned a lot about how wind works while working with his father to help the Tvind activists. He later decided to build a turbine to power his parents' farm:

My father and I made a small turbine, about one meter in diameter. It was basically just a rotor you could hold in your hand. And that rotor gave literally a lot of hands-on experience because when you do a thing like that, you really feel the effect of turbulence and changes in wind direction if you stand out in open air and let it spin in your hand. After that, I built a three-meter machine that I could mount on the wagon that we had for the tractor on the farm, so I could drive it out in the fields and test it and make sure how it would perform. Based on that, we decided to try and build a turbine to power the farm; I built a 15 kWh turbine with a ten-meter diameter and installed it in 1978.[22]

Stiesdal was very skilled at building small turbines and had a solid understanding of the technological challenges posed by wind, but he needed help with the metal work. He partnered with a local blacksmith and built a turbine that performed well. However, he was not ready to become an entrepreneur, so he sold his license to Vestas:

During the period I was building that, I met a fellow who was a blacksmith and who was very interested in wind energy. He was not an intellectual person, but he was very interested in energy and very good with machine tools, so he helped me with the mechanical machining of steel for my turbine. As a sort of reward for that, I figured out how to secure some research money to build a professional turbine. And I applied for a grant on behalf of both of us, and that grant was a kind of an inventors' grant and enabled him to build his first turbine in 1978. That one worked well, so he decided he wanted to build turbines and sell them to people. We sold two turbines in '79, but already I could see that I should not be part of a commercial arrangement with him because, even though I liked him a lot, it was also clear that he was not a good businessman, and I was going to start studying at university. He wanted me to join him and form a company, and I didn't want to do that. In the end I suggested that we find someone who could build our machine on license. So I got a hold of Vestas, which at that time was experimenting with Darrieus turbines and was not that successful, and I said to them, "Why don't you work with something that has more promise than the Darrieus machine?" So, in the summer of '79 we sold the license to our turbine to Vestas, and this is how Vestas got started in making proper wind turbines.[23]

Similarly to Erik Grove-Nielsen, Henrik Stiesdal was involved in OVE and greatly benefited from participating in the renewable energy meetings and fairs organized by this association during the late 1970s. In fact, the two of them met at a wind fair and collaborated on establishing the first safety rules for wind turbines in Denmark. As Stiesdal remembers,

During that period I had also become involved with OVE, and they arranged something called "wind fairs" where people would get together on a Saturday or a Sunday and simply discuss what they were interested in. I went to one, I'm pretty sure it was on the 23rd of

February, 1978. That was really interesting, and I gave my speech about my turbine, and there were others that gave speeches on what they were working on. Several were concerned about the safety aspects because we were concerned at that time about what would happen if turbines would fail and somebody got killed. Therefore, we sat down and picked a group, and one member of that group was Erik Grove-Nielsen. I think we were about six people in this safety group. And it worked for the next couple of years and ended up establishing the safety rules which you needed to follow to set up a wind turbine in Denmark. Even though it was sort of self-established, the safety rules were eventually taken over by the government when it introduced an approval scheme. This forum safety group and the wind fairs all meant that the heart core of the pioneers got to know each other. The meeting was a very good exercise for me personally, because I got involved with people who had the same interests and looked at some of the same issues as I did. So it was very interesting and very fruitful.[24]

In fact, it was because of his involvement in OVE that Stiesdal learned that Vestas, which had traditionally manufactured agricultural equipment, was also interested in manufacturing wind turbines; subsequently, he decided to sell them the license for his turbine. This is how Stiesdal recollects the event:

The way this thing [the Vestas license] worked was one of the fellows who was in this safety group was also a pilot. He was flying to the airport that is located near the Vestas factory one day, and he saw from the air that they had put up this Darrieus turbine. And I came to speak with him about other things later, and he said, "Are you aware that Vestas is also working on this?" I only knew them as a manufacturer of agricultural equipment, but they also turned out to be making hydraulic cranes, and they were making turbo-charged coolers for ships. So they kind of had three legs to stand on. Therefore, I sort of felt they would be likely to be open to a new idea: I gave them a ring and spoke with the owner, and then I met with him or rather the owner's son and the fellow that was managing director of the crane factory.[25]

Stiesdal is a modest person, but his influence on the wind energy industry has been enormous. During his career, he has been responsible for seventy-four inventions and eighty-five granted patents.[26] With the prototype developed by Stiesdal, the Danish company Vestas grew into the world's largest manufacturer of wind turbines. After completing his studies, Stiesdal first worked for Vestas, then for Bonus, which was acquired by Siemens in 2004. As Stiesdal says,

I had this very nice license agreement with Vestas, which essentially meant that I didn't have to work while I was at university. And then, after four years, my license expired. When it expired, I simply spoke with the fellow from Vestas who had bought the license; I had been in regular contact with him, so I simply asked him, "Can I get some work from you?" And then, in 1986, Vestas started getting into a crisis because their economy was not as good as it should be, and they had not been clever enough to foresee the American market development. I then left them, and I thought to myself, "I should concentrate on my university studies." But I couldn't live without the wind industry, so after a few months I thought, "I want to get back, I don't want to work for Vestas anymore. Who is the most serious competitor?" And Bonus was the cleverest, strongest, and highest quality competitor for Vestas. So I simply wrote them a letter saying, "Here I am; can I come and work for you?" I joined Bonus in early '87, and that's where I've been ever since. In '88, I became the technical manager of the company, and I stayed in that position until 2000. And then it had grown such that I felt my administrative burden had become too big; we made an arrangement and I became the technology manager so I could concentrate on technology.[27]

The Danish environmental movement contributed to the emergence and consolidation of wind turbine companies not only by producing activist-entrepreneurs and innovators. It also created a pro–renewable energy "spirit of the time" that attracted companies from other sectors into the market for wind turbines. Consider Henrik Stiesdal's account of the fact that Vestas and other companies began manufacturing wind turbines in 1980:

> Their motivation [Vestas], I think, was that they, like so many other companies in Denmark at this time, saw that there was a new market potential here that had not been realized previously. In a way it was strange, at least from today's perspective, that you could collect people from all over the country to speak about big ideas about the future energy supply. But it was all part of the spirit of the time. And that same spirit was also reflected in how companies saw opportunities, because they clearly saw opportunities at that time because of all the different types of concerns and voices. This was when we got in touch with Vestas, when the discussion about nuclear energy and safety aspects was at a maximum here in Denmark. At that time, people realized that "Oh, what the energy establishment believed would happen might not happen after all …" And that was what drove many of these companies, that renewable energy was simply growing in the public perception as something that had stopped being only for activists and was going to be something of a commercial nature.[28]

But how was it possible for the Danish environmental movement to shape the spirit of the time? To understand this process, it is necessary to examine the role of renewable energy advocates who formed "critical communities" within the Danish energy sector.[29] The environmental movement of the 1970s contributed to the emergence of renewable energy advocates, who identified the energy sector's problems and offered solutions that consisted of energy conservation and the promotion of renewable energy. The advocates worked to change the perception of renewable energy among energy professionals and entrepreneurs. Many began by criticizing the nuclear energy industry, and some were former nuclear scientists who experienced a conversion to renewable energy.

One of the most influential wind energy advocates in Denmark was Niels I. Meyer, a nuclear physics professor who became involved in both OOA and OVE. Professor Meyer was one of the authors of the alternative energy plan, which was published in 1976 and called for an energy sector without nuclear power and with a high contribution from renewable energy. Between 1971 and 1978, he was also the president of the Danish Academy of Technical Sciences (ATV), which published two reports proposing wind energy programs in Denmark, in 1975 and 1976. He was often a lightning rod for the anger of nuclear power supporters, who accused him of attempting to overthrow democracy in Denmark and to take Danish society back to the Stone Age. Some utility representatives even went as far as attempting to overthrow him as the president of ATV because he was promoting wind power and warning against the problems of nuclear power.[30] When the newly created Ministry of Energy published a second energy plan in 1981 calling for the introduction of nuclear power, Meyer was again one of the authors of an alternative energy plan. He worked closely with OOA and OVE to promote this alternative plan to Danish politicians and the general public.

Between 1982 and 1991, Meyer became the leader of the government Committee for Promoting Renewable Energy Systems, also known as the Steering Committee for Renewable Energy. In the mid-1980s, the committee had an annual budget of €4.6 million, and most of the budget was spent on development and demonstration of wind projects. Meyer tirelessly lobbied for more money, so the total funding for the Committee for Promoting Renewable Energy Systems reached approximately €30 million during its nine years of operation. Moreover, because the committee promoted new programs for offshore wind farms in the late 1980s, Denmark became the first country in the world to build large offshore farms.[31] As one of his former employees put it,

> Professor Niels Meyer was one of the first people to go against nuclear [power] in Denmark; he's a very brave man. He was very outspoken and got involved in the development of alternatives to nuclear power. He always managed to push hard for increased funds for renewables....People like Niels, who were lobbying politicians in government, were getting a lot of work done. They were not only working on energy planning, but also pushing politicians to give subsidies to alternatives. That's why, at the beginning, we could get 30 percent of the cost of a windmill from the state.[32]

Numerous other wind power advocates shaped energy professionals' perception of renewables and stimulated entrepreneurial activity in wind turbine manufacturing. Some were nuclear scientists working in the Risø Technical Institute; others were professors from the Danish Technical University. Interestingly, several energy experts have been secretive about the help they have given volunteer wind turbine builders, because these experts were working for institutions that supported nuclear power. Lars Albertsen, one of the key OVE organizers, describes the secret collaboration this way:

> Between 1974 and 1976, most people who got into wind were inventors and blacksmiths doing experiments with wind. They didn't know anything about it; there was very little literature about it. So we worked with some people from Risø and the Danish Technical University. At Tvind, they did not know how to build a large turbine, so they had to go to the Danish Technical University and to Risø. There were some engineers from Risø who helped them, but they were acting "in the dark." They didn't tell anybody that they were working for Tvind because they were supposed to be for nuclear power. Risø was the "bastion of nuclear power research" in Denmark, but some people weren't necessarily for nuclear. There was also a professor from the Technical University who helped them with the high-voltage wires.[33]

Advocates such as Lars Albertsen also played an important role in the exchange of information among wind turbine pioneers. Albertsen was a student when he became involved in OVE in the mid-1970s. He organized many OVE meetings and worked at the grassroots level for almost five years. As he emphasizes, the grassroots meetings were crucial for the rapid diffusion of technical knowledge:

> Some of the inventors in Denmark never talked to each other. That's where OVE made a big difference; probably the biggest difference. In the beginning of 1976, we organized a wind energy meeting where we got together and exchanged ideas and experiences. Small groups were getting together to solve technical problems in the design of the first turbines. After a while, we had groups working on different things. The technical meetings had a very big role in the first few years. Everybody who worked with wind energy had to go to those. If you didn't, you couldn't get the latest information. Some people wanted to work by themselves, and they couldn't; they found out that it was better to collaborate. That was our policy: everything that we did and our meetings had to be open to everyone else.[34]

In 1980, Albertsen left OVE and was employed by Niels Meyer to work for the Committee for Promoting Renewable Energy Systems. He continued to be an important wind power advocate; in fact, without support from the Committee for Promoting Renewable Energy Systems, numerous Danish wind power entrepreneurs would have failed. The committee's financial support was critical for renewable energy entrepreneurs in the early 1980s, when the wind energy industry was in its infancy. Meyer, Albertsen, and others involved in the committee worked hard to gather financial resources and to achieve independence from the Ministry of Energy. As Albertsen emphasizes,

Niels was very smart; he got a lot of money from the government. We were sitting with a lot of money, but we didn't want to do research: we decided to promote renewables. The guidelines for our work were very liberal, so we could do as we wanted. We could give money to people that nobody else wanted to give to because we believed in them. We could do this because the committee was part of the Ministry of Industry, which had no nuclear people—like the Ministry of Energy did.[35]

Perhaps the most important wind power advocate in Denmark was Preben Maegaard. Like many other Danish wind power advocates, he became involved first in the antinuclear movement and then in renewable energy. This is how he describes his decision to become a wind power pioneer:

In the beginning, I wrote up some long articles on nuclear energy. My special interest was to document the nuclear power industry, and it was my contribution to the discussion on nuclear power. Many of the nuclear power plants were stopped for months because the availability factors were highly overestimated. Official figures came from German and American power plants, but those figures were not real. We asked, "If Denmark got one of those, what kind of supply would we be getting into?" That was my contribution to the debate. I live in a rural area of Denmark and when I asked people around there, they weren't interested in nuclear energy. They asked for biomass, wind, solar, or something similar that could work well. This was the way that I got involved in nuclear energy. In 1974, I asked myself, "Why couldn't we supply ourselves with wind and solar?"[36]

Like other wind power pioneers, Preben Maegaard emphasizes the crucial role played by OVE and the "Saturday meetings." Maegaard argues that in the mid-1970s, it was imperative for environmental activists to change their strategy from saying no to nuclear power to saying yes to renewable energy:

The technologies were not available for renewables; there was nothing to get us there. So OVE had to focus on how to develop these technologies with realistic size, cheaply and reliably. It was basic research work. There were no university people focusing on renewable energy solutions, there was no research. Even in 2002, only 8 percent of OECD [Organization for Economic Co-operation and Development] countries' energy R & D expenditures were for renewables, while 60 percent went to nuclear and 20 percent to fossil fuels. This sector has had a very low priority for governments and universities. In OVE, we had to prove practical results. The organization said no to something, but had to say yes to something else as well. We had to develop the technologies—it wasn't just writing an article at the university. So, many individuals and small companies were working on different ways of designing wind, solar, biogas, and so on. Eventually we had meetings organized on Saturdays. I was coordinator of these meetings for a number of years.[37]

The meetings organized during the 1970s by Maegaard were crucial for the exchange of information between wind power innovators. As he emphasizes, the free exchange of ideas made possible the gradual improvement of wind turbine technology:

> The main way of exchanging know-how about wind energy wasn't through professional or university conferences. It was through these Saturday meetings with interested and involved citizens. There was nobody trying to create patents. There is the idea that a patent causes progress and development, but here we had the opposite viewpoint. We had a very cooperative movement.... When you develop new technology, you don't know what's going to be successful. It's like when you develop medicine, you have to try many different kinds because you don't know which one will be most effective. When we started in 1975–76, we didn't know which wind concept was going to achieve reliable and safe results. You can have windmills with vertical and horizontal axes, one, two, or three blades; you can have blades made out of steel or fiberglass, etc. There are some who think that some combinations would be the best and others who think that others would be the best. When you are in 1976 and look into the future, you cannot know which will be the leading concept. The vertical axis seemed to be the best model for wind turbines, but they didn't work in the end. Why did they work on concepts that later proved to be the wrong ones? It was the testing that brought about the right concepts.[38]

Maegaard encouraged wind pioneers to build turbines using components from different suppliers. In 1976, during one of the first OVE meetings on wind energy, he introduced the concept of the component wind turbine. He explains the concept this way:

> An important question was whether we should really build all the components.... If you look at a bicycle, it's basically a frame; there are also the spokes, the chains, the tires, the pedals, etc. Everything comes from specialized suppliers. If you buy a frame, you can go buy your other components and make an excellent bike. The same concept is applied to the wind energy industry. We made it so that everyone who had a workshop could create a wind turbine. Blacksmiths took the blade, the control system, bought a gearbox from Belgium, got a tower from the tower factory, and put it together. And it worked well.[39]

This simple concept had very important consequences for the way in which the Danish wind energy industry developed; as Maegaard argues:

> The structural division of a windmill into individual components was the innovation which would give Denmark decisive comparative advantages as regards the windmill production. It was far more manageable and much cheaper to buy the components than having to develop and test all the parts on your own, or to construct and build production plants for all the components, including the wings. This is

mainly true as regards the many small companies, but also big companies failed trying to do it all by themselves.[40]

A number of wind groups were formed during one of the meetings organized by Maegaard at the end of 1976; among those was one group working on developing fiberglass blades. At this meeting, Maegaard suggested that a number of wind turbine pioneers join in and invest in building a mold for fiberglass. One of the participants in the meeting volunteered to do calculations for the blade, and a few others—from Tvind—volunteered to build it. Later, Grove-Nielsen bought the mold and launched his company, Økær Vind Energi. Moreover, when Grove-Nielsen had problems launching his company, Maegaard got in contact with him and, through the Northwest Jutland Institute for Renewable Energy, brokered a contract for a number of blades.

The meetings were also instrumental in attracting numerous companies to wind turbine manufacturing. Many representatives of small companies, particularly those involved in manufacturing agricultural equipment, participated in the OVE meetings and benefited from the free exchange of information. As Maegaard points out, Vestas was one of them:

> The Danish Blacksmith Association came to one of the OVE meetings in 1978—the chairman and vice chairman were there. It was a meeting where there were a lot of people. I went up to them and asked, "Why are you here?" They said, "We represent two thousand small companies with over twenty thousand people employed. We have a tradition to make designs available for our members because we don't have many engineers. We thought these renewable energies could be of interest to our members in the future." Vestas was actually part of that association, as it was a small company that manufactured agricultural designs using blacksmith practices. This was the first time that industrialists showed interest in renewable energy in Denmark. They took this interest to politicians and asked for a good policy framework to promote renewable energies. Politicians tend to listen to industry more than to grassroots, because they employ people. So, now renewable energies got to the political level.[41]

In 1979, Maegaard became the chairman of OVE, a position he maintained until 1984. During this time, he continued to help wind energy entrepreneurs by organizing meetings and by offering free consultations and technical training sessions. In the early1980s, he was also a consultant for Bonus, a company that had been previously building irrigation equipment but, like many other companies, perceived an opportunity to learn from renewable energy pioneers and build wind turbines. As he explains:

> I was invited by the owner of Bonus there. He wanted me to speak about the opportunities with renewable energy. He said he had some thoughts on how to build windmills and they had put something together. By that time, I had 3–4 years of experience in the design of

wind turbines, especially in blacksmith practices. Bonus wanted my advice on the main criteria to make a good wind turbine with a reliable concept. I told them what to do and what not to do. In terms of marketing, there were some companies that wanted to put wind turbines all over the country. They wanted to become well-known. I told them that it was new technology, and that they should not install wind turbines more than one hour away from their facility because it takes a lot of effort to maintain them. I also gave them some information about the factor levels of the gearbox, which makes a difference in terms of whether it will last in the long run. I worked with them only for one or two years because I was not interested in being employed by a company.[42]

In addition to his work with OVE, between 1974 and 1979 Preben Maegaard also led the Northwest Jutland Institute for Renewable Energy (NIVE), which included renewable energy enthusiasts in various professions: teachers at the technical schools and high schools, engineers, farmers, and blacksmiths. In 1979, NIVE decided to create a Folkecenter [People's Center] with the purpose of training individuals interested in developing a career in wind or other renewable energy industries. With support from Niels I. Meyer and Lars Albertsen—from the Committee for Promoting Renewable Energy Systems—as well as from key political allies from the Social Democratic Party, the proposal for creating the Nordic Folkecenter for Renewable Energy (NFRE) was approved by the Danish parliament in 1983.

Maegaard has been the director of the NFRE since 1984. In this capacity, he has been responsible for the technological innovation of various energy systems, including the design, construction, and implementation of wind turbines ranging in size from 20 to 525 kW. The NFRE has contributed to entrepreneurial activity in the wind energy industry, and was particularly active during the 1980s. One of its functions was to provide expertise to wind pioneers who wanted to take advantage of the California wind rush but did not have enough technological resources. For example, engineers and blacksmiths working for the Folkecenter developed the design for the *sme-demester* [blacksmith mill], which was offered free of charge to anyone interested in manufacturing it. Many experts argue that the blacksmith mill has not been a commercial success—it is estimated that, by 1990, only 185 of those turbines had been built.[43] Moreover, only one company using this design (Lolland) actually sold turbines on the California market.[44]

The Folkecenter has, nevertheless, played an important role in entrepreneurial activity by functioning as a training ground for wind turbine innovators. For example, engineer Knud Buhl Nielsen, the main designer of the blacksmith mill, later become the well-known designer of the Nordex turbines in Denmark and later in Germany.[45] As Maegaard points out, the Folkecenter provided employment for engineers and technicians who later joined the wind energy industry. The center was supposed to "create employees for corporations and organizations, which has indeed been the case to a large degree. Many people in the industry and other places were first introduced to working with renewable energy at the People's Center. This is where they became professional and from where they could go on."[46]

Additionally, the Folkecenter helped wind energy entrepreneurs who did not have the technical resources to start up their own companies. According to Maegaard,

> We had six or eight engineers employed, doing design and construction work for small companies that wanted to take opportunity of the California market. We started with small turbines at the beginning of the California market, and then we went to large, for the time, 500 kW turbines. Basically, we did design work for companies that wanted to go into manufacturing and that did not have their own skills and experience. We gave them some basic design, and then they built prototypes. After some time, they employed engineers, and we trained them and they went on their own. For example, one of those companies was Dencon; another one was Vindsyssel. We stopped this in 1992 because we could see the industry was so strong that it made no sense to continue.[47]

Perhaps even more important is the fact that, through the NFRE, Maegaard has contributed to the setup of numerous small-scale pilot projects worldwide. As part of the Folkecenter's activities, numerous renewable energy enthusiasts from all continents have spent between three and nine months as trainees, and "have obtained valuable hands-on experiences that paved the way for dedicated and leading positions in their professional careers."[48] Under Maegaard's leadership, the NFRE launched projects in cooperation with local companies, NGOs, and governmental authorities from Eastern Europe, Asia, Africa, and North and South America. Indeed, Maegaard's vision is that renewable energy know-how should be "spread to several corners of the globe, to the benefits of the world society, and future generations."[49]

Last but not least, it is important to emphasize that Preben Maegaard and other people involved in the NFRE contributed to the transfer of wind energy technology to both developed and developing countries. Consider the important case of technology transfer from the NFRE to General Electric. After the collapse of the California market, in 1988–1989 the NFRE applied for and obtained a grant from the European Union to develop wind turbines with a minimum 500 kW capacity. Because one of the requirements of the grant was to have foreign partners, Maegaard contacted a well-established gearbox manufacturer in Germany. As he recollects,

> We knew that Tacke was a well-known gear manufacturer. They made heavy-duty gearboxes for marine applications, so we developed two prototypes with them. The Folkecenter had two-thirds of the project, and Tacke had one-third. We agreed to have identical solutions for

nacelles, blades, and towers. Most of the design was done here; so, this was Danish know-how transfer to Germany. Then, the prototypes went up and they both worked very well. Later on Tacke made bigger (600 kW) versions—this is how Tacke went into this business. Tacke was later taken over by MAN, which makes trucks and a lot of other things. But a separate division was created with the name Tacke Wind, which later got into some trouble, was taken over by Enron, which was then taken over by GE Wind. That's why I say that GE turbines have their roots in the Folkecenter design—but don't think you will ever make them confess this! GE Wind says they had their own turbine design, and they will never confess where it came from.[50]

Consider also the case of technology transfer from the NFRE to Suzlon, a large Indian wind turbine manufacturing company. According to Maegaard, "We gave one of our designs to another company called Vindsyssel, which passed it on to a company in Germany called Süedwind. Then, when Süedwind was merged with Nordex, they founded a design office that worked, among others, for Suzlon. This office was later taken over by Suzlon. This is just one story, but I think we can trace other similar ones about the impact of the Folkecenter windmill design."[51]

As well as being a tireless advocate for wind power and other renewables, Maegaard was also a senior vice president of Eurosolar between 1991 and 2006. In 2001, he became the first president of the World Wind Energy Association (WWEA), a position he held until 2005. In 2006, he became the first president of the World Wind Energy Institute (WWEI), an organization involving institutes from China, Brazil, Cuba, Canada, Russia, Egypt, and Denmark. In recognition for his life-long career as a wind and renewable energy advocate, he has received numerous awards, including the Organization for Renewable Energy 1978 Solar Prize; the Association of Danish Engineers 1987 Environmental Prize; the Denmark's Wind Mill Owners Association 1992 Wind Energy Prize; the 1997 Eurosolar Prize; the 2002 GAIA Prize; and the 2005 Nuclear-Free Future Solutions Award.[52] In 2008, he received the World Wind Energy Award, as recognition for the fact that "his special dedication has always been to come to a more democratic structure of the energy supply and he has supported community power approaches as a powerful tool for that goal."[53]

Environmental Activism and Wind Turbine Manufacturing in Germany and the United States

The environmental movement also contributed to the emergence of wind power entrepreneurs, innovators, advocates, and champions in other countries. In Germany, for example, one of the most influential entrepreneurs and innovators has been Aloys Wobben. As an engineering graduate student, he was firmly opposed to nuclear power and became interested in developing alternative energy. He built his own small (5.5 kW) turbine in 1975; in 1984 he began building a 22 kW turbine inspired by the Danish concept and founded the company Enercon.

Like many other activist-entrepreneurs, Wobben has been a driving force for the implementation of new technical innovations. In 1985, Enercon was already producing a variable-speed 55 kW turbine—the first professional turbine with variable speed ever installed. In 1992, Enercon figured out how to build windmills that were gearless and functioned without hydraulics. Enercon has also produced some of the largest turbines in the world; for example, in 2002 it introduced the E112 model, which could produce 4.5 MW and held the record for the largest turbine in the world for two years. By the end of 2008, Enercon had installed more than thirteen thousand wind turbines and had production facilities in Germany, Sweden, Brazil, India, Turkey, and Portugal. Wobben has remained Enercon's chairman and managing director since he founded the company. He is also a well-known environmentalist who has received numerous awards; for example, in 2000 he was given the German Environment Award by the German Environment Foundation, and in 2004 he received the Eurosolar Special Prize for Extraordinary Individual Commitment.[54]

In the United States, many of the wind energy entrepreneurs, innovators, and advocates during the 1970s and 1980s were also environmental movement activists and sympathizers. Indeed, as one newspaper article noted, "Wind industry guys are the straightest-shooting people. Most got into it because they had an environmental ethic."[55] As in the case of Denmark, some of the wind power entrepreneurs and advocates were nuclear scientists or engineers who, under the environmental movement's influence, decided to dedicate their careers to renewable energy research and advocacy. For example, William Heronemus was an American nuclear engineer who became an environmentalist and a wind energy pioneer during the 1970s. His assiduous activities as a writer of articles related to wind power, and as a teacher about wind and other renewable sources of energy, encouraged many of his engineering students to become wind entrepreneurs. As one author notes, "The 50 or more students who graduated from his programs quite literally became the modern wind industry as they started up companies, worked in federal laboratories, or filled other positions in the private and public sectors."[56]

One of the most influential wind power entrepreneurs in the United States was James Dehlsen. Although he was not directly involved in the anti-nuclear or environmental activism of the 1970s, Dehlsen considers that he was influenced by the environmental movement of the 1970s and, in particular, by Amory Lovins. He remembers that reading Lovins's book *Soft Energy Paths* made a big impact on him and motivated him to enter the renewable energy business. As he recollects,

I think I've always been pretty environmentally oriented. Growing up in Latin America, I spent most of my time outdoors....In the mid-70s

I had developed a product called Triflon, a fluid lubricant based on micron-sized Teflon particles, and quite an effective product. That was during the energy crunch in the 1970s, so I had a product that had the benefit of reducing friction and basically making machines last longer. But in the process of all that, I became more interested in energy and wound up reading Amory Lovins's book... After that I really wanted to pursue renewable energy, and I sold Triflon in 1980 and used the proceeds to form my company. And that's essentially at the beginning of wind power.[57]

In 1980, Dehlsen formed Zond Systems, a company that was involved in designing and manufacturing wind turbines. Soon afterward, however, Dehlsen realized that the Danish wind turbines were technologically superior and decided to order wind turbines from Vestas. What started as a small order rapidly grew because of the California investment tax credit: from 150 turbines in 1981 to 1,100 turbines in 1985. As Dehlsen says, "I ended up buying basically all of their [Vestas's] output for the first period during the 1980s. I bought about three thousand machines. That's basically how the Danish industry got launched. It was really very much the development side of the activity in California that essentially created the Danish business."[58]

Dehlsen's company not only contributed to the growth of the Danish wind turbine manufacturers; it also launched innovative programs with U.S. utilities. When the tax credits expired, Zond survived the crash of the California market by reengineering older wind projects it had built. By the late 1980s, however, Dehlsen had convinced Florida Power and Light to invest in "the first significant institutional financing in the industry," leading to Zond's 77 MW, $157 million Sky River facility in 1990. As Dehlsen proudly maintains,

> The project was realized despite what "experts" indicated were insurmountable obstacles, including Southern California Edison's (SCE's) extraordinary requirement that we provide 75 miles of 220 KV transmission lines, adding $30 million to the project cost. This, however, was the largest project in the industry to date, and its accomplishment became widely recognized for its performance and as a new model for wind projects of substantial scale and remote from the existing grid.[59]

Following the success of the Sky River project, Dehlsen also got interested in manufacturing wind turbines. Realizing that future growth could come only if Zond built its own turbines (thereby removing the manufacturers' profit margin), Dehlsen applied for and was awarded a grant for turbine development by the National Renewable Energy Laboratory and the U.S. Department of Energy. In 1993, Zond developed a 550 kW turbine called the Z40 and, with research and development grants, built several demonstration projects in 1995. The next year, Zond acquired patent rights from Kenetech, a wind industry competitor that had declared bankruptcy. In 1997, Zond began work on a variable-speed 1.5 MW wind turbine. In anticipation of the company's capital needs for manufacturing, Dehlsen partially sold the company to Enron in 1997. At Dehlsen's urging, Enron acquired Tacke, a German wind turbine manufacturer that had fallen into bankruptcy. As Dehlsen points out,

> By combining the best technologies from Zond and Tacke, the TZ 1.5 MW turbine went into production and has become one of the leading turbines in the global market. While our wind energy business had stellar performance under the Enron umbrella, Enron collapsed into bankruptcy. In 2002, General Electric purchased the wind turbine technology rights and manufacturing assets, and the newly formed GE Wind Division advanced rapidly with the 1.5 MW turbine to become the third largest turbine manufacturer in the industry.[60]

After he sold Zond, Dehlsen became a wind power entrepreneur and innovator once again. He formed Clipper Windpower with his son Brent and, with support from the U.S. Department of Energy, designed a new type of drivetrain, launching the 2.5 MW Liberty wind turbine. In April 2008, Clipper signed an agreement allowing the Crown Estate of the United Kingdom to purchase Clipper's prototype of the world's largest offshore wind turbine, Clipper's 7.5 MW wind turbine, also known as the Britannia Project. In 2007, Clipper was recognized by the U.S. Department of Energy with an Outstanding Research and Development Partnership Award for its "outstanding contribution toward industry advancements." This award recognizes the Liberty wind turbine for attaining "unparalleled levels of efficiency and reliability and reduced cost of energy."[61] In October 2008, Clipper formed a fifty-fifty joint venture with BP Alternative Energy to develop the Titan wind project, a 5.05 GW wind energy development located in South Dakota—a project expected to be the world's largest wind farm. For his lifelong accomplishments, Dehlsen was inducted into America's National Environmental Hall of Fame as the "Father of American Wind Energy" during a formal ceremony at the end of 2008.[62]

Yet another influential wind energy advocate in the United States is Paul Gipe.[63] As a student during the early 1970s, Gipe became the chairman of a very active environmental group at Ball State University in Indiana. He became interested in renewable energy because he wished to limit the environmental effects of conventional energy sources, particularly those of coal and nuclear power. He contributed to the passage of the National Surface Mining Act, which regulates the strip-mining of coal in the United States, and coauthored a study with the title "Surface Mining, Energy, and the Environment."[64] In 1976, he became interested in wind turbines because "solar panels are boring, because they don't do anything—they just sit there. But wind turbines, of course, fascinate people, particularly males, because they go around."[65] From 1976 to 1984, he was a consultant with an emphasis on technical issues and environmental-impact analysis of wind turbines.

Paul Gipe was also an assiduous promoter of wind energy, and he provided seminars, workshops, and training aid. Between 1984 and 1985, he worked as director of corporate communications for Zond Systems, one of the largest wind turbine manufacturing companies in the United States at the time. Between 1985 and 2004, he launched a company specializing in evaluating wind turbine technology and reporting on developments in the wind industry for industry associations (the American Wind Energy Association and the British Wind Energy Association, for example), governmental organizations (such as the U.S. Department of Energy and the National Renewable Energy Laboratory), and nonprofit organizations (including the Izaak Walton League and the Sierra Club).[66] From 1985 to 1992, he served as West Coast representative for the American Wind Energy Association, and in this role he "represented them at public meetings…and promoted the interests of the wind industry."[67] Additionally, Gipe is the author of several basic books on wind energy, which have contributed to the dissemination of wind power technology to a broad audience, and he is the main organizer of the campaign for feed-in tariffs (FITS; advanced renewable tariffs) in North America. In 2008, he received the World Wind Energy Award, together with Jane Kruse and Preben Maegaard, in recognition of the fact that he is "the most important advocate for wind energy and community power in North America."[68]

It is undeniable that environmental activists made a major contribution to wind turbine manufacturing, particularly in the early years of the industry, by becoming wind power entrepreneurs, innovators, and advocates. But environmental activists and sympathizers also contributed to the growth of the industry by becoming champions of wind turbine technology while working for large energy companies. Consider the case of James Lyons, who worked as an engineer for General Electric between 1970 and 2008. Although he does not consider himself an environmental activist, he strongly believes in the environmental movement's core principles. As he says, "I spent a lot of time outdoors, canoeing and hiking and doing things, so I feel I have a strong environmental ethic. I think the environmental movement of the '70s, although well-intentioned, was not very scientifically based. So I'm not going to align myself with them, but I definitely believe in the core principles of what they were after and I think it's ridiculous to burn coal if you can have free energy here that is perfectly clean."[69]

Lyons became interested in wind energy during the late 1970s, and pursued a PhD on variable-speed wind turbines. After he received his doctorate, he returned to General Electric; in 1989 he joined the GE research unit, and in 1999 he became the chief engineer for electrical and electronic systems. Since the early 1990s, Lyons has been not only an innovator who was awarded twenty-eight patents but also "a corporate champion for renewable energy within GE and one of the founding leaders of GE's wind energy business."[70] In the mid-1990s he decided "to work for a time trying to understand the state of the art wind industry and see if we could work as a component supplier."[71]

Although GE did not become a supplier of wind turbine components at that time, Lyons was convinced that wind power was a very promising industry, and he decided to start an internal campaign to get the company into wind turbine manufacturing. As he remembers,

In 1998 (I was already appointed chief engineer at that point), myself and another gentleman decided to start a campaign to get GE in the wind business with both feet. So we started a campaign building the business case for wind as an attractive new business for GE. That was a very fruitful period, as we built internal support. We actually brought that case to Jack Welch [former CEO of GE] twice, actually twice to Jeff Immelt [former CEO of GE] too. We worked on that for, I would say, three and a half years before getting approval to enter the wind industry.[72]

Lyons concentrated on making a strong business case for wind power and on finding influential allies within the organization. Here is how he describes his efforts:

We started building the case, expanding the size of the team to include strategic marketing at the GE power systems business, the business development group within power systems, both in the U.S. and Europe....One important ally was Frank Blake, who ran the business development activities within GE power systems. He got it, if you will, what the potential was, whereas a lot of the GE folks were saying, "These are wimpy turbines. We make 500 MW gas turbines." But he got it; he understood that the world was shifting, whereas a lot of others saw the wind business as a bit flaky, California kind of stuff. So we had to make a very strong business case, we didn't make an environmental case. Blake gave us free run in the business development community to go talk to these companies that could become potential acquisition targets. Another team member that worked for him was Swedish and, being a European and seeing what was going on in Germany and Denmark at this time, we kept telling the GE guys, "You guys got to get out of South Carolina and Schenectady, New York, and get over to Europe and see what's going on." Another one of the guys instrumental in getting us started was the strategic marketing leader for GE power systems. He was very valuable in that he was experienced at his job and he went around and set up these meetings and talked to a lot of the customers. So, it wasn't just a research guy telling everybody the world is shifting, but you had the strategic marketing director saying the same thing.[73]

Lyons's sustained efforts paid off when, in 2002, GE's CEO decided to enter the wind business. Rather than designing wind turbines "from scratch," GE acquired Enron Wind because it possessed considerable know-how from the acquisition of Zond and Tacke. Lyons acknowledges the fact that the GE turbines are descendents of German turbines and argues that, despite the skepticism of many GE employees, the wind turbine business is one of the best deals GE has ever made:

We got thumbs-up, if you will, from Jeff Immelt to enter the business. At that point, we presented three different scenarios or options on how to do it, and then Enron went bankrupt two months later, which made our choice pretty much self-evident. We acquired a number of assets: the former Zond in Tehachapi, California; Tacke Wind Energy in Salzburg, Germany; Airpack in the Netherlands—a blade manufacturer; and a turbine assembly plant in Spain. Most of the engineering was in Salzburg, Germany. The GE 1.5 MW is essentially derived from the Tacke Wind Energy 1.5 MW turbine. GE has just installed its 10,000th unit, which is just fabulous, a revolutionary product....GE sold six and a half billion dollars worth of wind turbines last year, so it was GE's best start-up ever. But many people at GE still don't grasp that this is a fundamental generation that it's going to be bigger than nuclear and as big as the gas turbine business. It's really hard for the thousands and thousands of people that grew up with thermal generation to understand that this is here to stay.[74]

Finally, it is important to note that although the U.S. environmental movement played a role in the emergence of wind turbine entrepreneurs, innovators, advocates, and champions, fewer of the wind energy entrepreneurs in the United States have been motivated by an environmentalist ethic than in Denmark or Germany. As one author observes, the California investment tax credits in the early 1980s were "America's last great tax shelter," which made wind farms "the darling of Wall Street and of those who make their living counting and hiding other people's money."[75] And another author observes that "wind energy in the United States had a less pronounced social-movement side [than in Denmark]. In the wind industry the countercultural element was only one stand among others, including entrepreneurs, corporate energy firms, former military engineers, and Wall Street investors."[76]

Environmental Activism and the Development and Operation of Wind Farms

At the beginning of the 1990s, the major electricity companies produced most of their electricity from fossil fuels, nuclear power, or a combination of both. While wind and other renewable sources were also used to produce electricity, the developers and operators of renewable energy projects were mostly small, independent power producers. Large electric utilities sometimes purchased renewable energy from independent providers but, with a few exceptions, did not develop or operate wind farms.[77] Gradually, however, more and more electricity companies started to develop or operate large wind power projects, or do both. In the United States by the end of 2008, NextEra Energy (a subsidiary of Florida Power and Light) had about 6.3 GW of wind power in commercial operation, while Xcel Energy and

Duke Energy had approximately 2.7 GW and 500 MW respectively. In Europe, a growing number of electricity companies such as Npower, E.ON, DONG, and Vattenfall operate large wind farms and have ambitious plans to build even larger ones, either on their own or in partnerships.[78] Additionally, companies that specialize in the development of new wind power projects— such as Ecotricity or Community Energy—have also emerged as important players in the energy sector.[79]

The environmental movement is a major force that "stressed" the electric utility system and "changed the world in which utility managers operated."[80] Starting in the 1970s, environmental groups and activists highlighted the crucial relationship between energy use and environmental damage, and challenged electricity companies' models of growth. Over time, the number of interactions between environmental organizations and electric utilities increased; some of the interactions were antagonistic, others were collaborative. Environmental activists in Europe formed independent cooperatives and companies that specialized in developing and operating wind farms. In the United States, environmental groups influenced electricity companies' decisions to invest in wind power mainly through protests, litigation, lobbying, joint marketing, and counteracting local opposition to wind farms.

Environmental Activism and the Development and Operation of Wind Farms in Europe

Since the 1970s, Danish environmental activists and wind turbine owners have pressured electric utilities to accept interconnection of wind turbines and to offer favorable feed-in prices for electricity produced by wind cooperatives. Since the 1980s, growing numbers of Danes have also bought shares in wind turbines and formed cooperatives. Many of those wind turbine cooperatives were formed by environmental activists driven by environmental concerns more than by economic self-interest. For example, the Helligsø windmill cooperative was formed in 1988 even though it was a financially insecure enterprise because "the driving force was not a dream of economic gain.... what [people] wanted was to produce pollution-free energy. According to calculations, a windmill could be called 'pollution-free' when it had operated for one year in the sense that the energy production had by then made up for the consumption of resources necessary for the building of the mill."[81]

For many environmental activists in Europe, becoming independent power producers by joining wind turbine cooperatives was an opportunity to opt out of the dirty electricity sector. Grassroots environmental activism during the 1970s and early 1980s contributed to the rapid growth of the wind turbine cooperatives in some countries.[82] About 33 percent of all wind capacity in Germany has been built by associations of local landowners and nearby residents. In Denmark, about 25 percent of the wind generating capacity has been developed by wind turbine guilds or cooperatives.[83]

Between the late 1970s and early 1990s, voluntary power purchase agreements were set up and renegotiated between utilities and independent wind power producers. The relationship between these independent producers and electric utilities, however, was often contentious. Because utilities had little experience with small-scale, dispersed systems such as wind turbines, "most Danish electric utilities were highly skeptical about wind power, and they were not interested in offering favorable feed-in (payback) prices for wind electricity."[84] Disagreements over grid connection for new wind turbines led to the breakdown of voluntary agreements in 1992 and to the intervention of the government, which set up a FIT at 85 percent of the utility production and distribution cost.

In addition to joining wind turbine cooperatives, environmental activists also became wind-farm entrepreneurs. This happened predominantly in the United Kingdom, which deregulated its electricity markets during the 1990s. Consider the case of Dale Vince, a British environmental activist who founded Ecotricity—the largest company specializing in wind-farm development in the United Kingdom. Although he did not have a technical background, like some of the wind turbine entrepreneurs in Denmark or Germany, Vince became interested in developing small-scale wind farms because he shared the wind turbine entrepreneurs' deep passion for clean energy. After spending about ten years "living 'on the road' and searching for an alternative way to live," he was by the early 1990s living in an ex-military vehicle and using a small wind turbine to produce his own electricity.[85] He decided "to bring change to the electricity industry" by building a large wind turbine—thus, after years of battling "bigots, planners, and big power companies," he finished building an industrial-size wind turbine in 1996.

While fighting to build his first large wind turbine, Vince realized that the main obstacle for developers of new wind farms was getting a fair price for their electricity; he decided to found Ecotricity to address this problem.

> I wanted to change the world in some way; I thought I could do more by dropping-in than by dropping-out. I've been an environmentally concerned person all my life; dropping-out was all about environmental concerns, concerns about the sustainability of modern life.... It took about five years to get the first wind turbine built, and it was along the way that I realized the big obstacle was getting a fair price for the power. This was before green electricity existed in the world as a product. I went to see the local power company, and I asked them if they wanted to buy green electricity and they laughed. They said, "Nobody wants green electricity, what is it, anyway?" and they gave me a rubbish price. So, I left that meeting having decided that I needed to get directly to the end user in order to get a fair price. I got a supplier license, since in the newly liberated market it was possible to become an electricity company, and founded Ecotricity in 1995—it was about getting a fair price so that I could build more wind energy, and building wind energy was about saving the world from climate change.[86]

According to Vince, Ecotricity's success is due to growing public concern about global climate change and to a business model that is fundamentally different from that of electric utilities.

> We conducted a survey recently and 98 percent of our customers said they joined Ecotricity in order to do something about climate change. We've grown over the last three years mainly because of growing awareness about climate change; I think you can see this everywhere in the last three years, that people start looking for solutions, which is what we do....Our approach is fundamentally different from that of big power companies: we don't exist to serve shareholders, we exist to build windmills and make the world a better place. Because we don't answer to investors and shareholders, we are able to put all our money back into what we do—that explains why we can spend so much more money on wind farms than they can, because they have to give returns to people. They are in this to make money; we are in this to make windmills.[87]

Besides having a philosophy that is different from that of conventional utilities, Ecotricity also uses an innovative business model. Through its Merchant Wind Power scheme, Ecotricity builds, owns, operates, and maintains the turbines—in return, companies buy the green electricity and gain, as Vince puts it, "a stunning environmental signpost." Vince emphasizes that Ecotricity was able to attract a number of large companies (Ford, Michelin, Prudential, and others) because it uses an original business model that makes wind power competitive with traditional energy:

> We decided that wind energy could and should be made to work without government support. So, we came up with the idea of deeply embedded wind assets on the customer side of the meter and the grid connection—this way we save the cost of delivering the power, which is often a third of the electricity bill. You trade off for low wind, because industrial sites are not usually very windy, but you gain on the transmission cost. The big pitch for commercial customers is that we do all the work for feasibility and planning, we finance and build the wind turbines, and they just buy the power when the wind blows. And when they buy the power, they are doing it at a competitive price. So, they get the big green symbol and they get green electricity, but they don't have to lift a finger. That's why lately we've been able to sell for less than the cost of conventional energy.[88]

British environmental activists also have had a direct influence on a few electricity companies' decisions to invest in wind farms. Consider the case of a collaboration between Greenpeace and Npower, which resulted in the construction of an offshore wind farm and in the marketing of a green-power program named Juice. According to Greenpeace UK's former director, Stephen Tindale,

> Juice was innovative in the sense that it was the first no-premium green program. Greenpeace campaigned in favor of the North Hoyle

wind farm. In the U.K., planning is the biggest constraint on renewables, so it was quite important for Npower to have Greenpeace supporting their proposed wind farm. We also did some publicity around Juice....There are two reasons why Juice was successful: one is because it was a no-premium program; another one is because Greenpeace was supporting it. So, both in terms of getting planning permission and in terms of getting to their customer numbers quickly, Greenpeace was significant.[89]

Environmental Activism and the Development and Operation of Wind Farms in the United States

While in Europe environmental activists changed the electric utility system mainly by forming independent cooperatives and becoming wind-farm developers, in the United States their influence was felt mainly through protests, litigation, lobbying, joint marketing, and efforts to counteract local opposition to wind farms. Because American environmental groups had relatively little influence on the federal energy policymaking process, many of them concentrated on persuading utilities to invest in renewable energy. As early as the mid-1970s, environmental groups such as the Natural Resources Defense Council were suing utilities for planning to build new power plants without considering alternatives such as conservation measures.[90] During the 1980s, numerous American environmental groups advocated for "demand-side management" as a solution for the projected increase in electricity consumption.

The first "collaborative"—a plan for energy-efficiency and demand-side management programs developed in collaboration between environmental groups and electric utilities—was set up in 1988.[91] By the end of 1991, over twenty-four utilities in ten states had worked with environmental groups to reduce energy consumption through demand-side management programs.[92] Therefore, as Hirsh (1999, 221) notes,

> The prominence of environmentalists as the driving force behind the collaboratives suggested that this group of stakeholders had gained elevated standing. Contesting traditional utility strategies for more than two decades, environmental advocates effectively cultivated public opinion and exploited provisions of new legislation. They also convinced regulators and power company executives that conservation policies could reduce construction expenditures and allow utilities to earn handsome profits.

The contacts between U.S. environmental groups and utilities multiplied during the 1990s. The rise of the global climate change issue to the top of the list of environmental problems during the second half of the decade led numerous environmental groups to adopt a more confrontational stance toward utilities. Because the global climate change problem is so large, environmental groups and activists argued that conservation measures are not sufficient, and they pressured utilities to invest in renewables. Many environmental organizations used confrontational tactics such as protests, lawsuits, and lobbying for climate change regulation, while others used cooperative strategies such as joint marketing of green-power programs. When they dedicated significant resources to clean energy campaigns and when they operated in a favorable social context, environmental groups had a significant influence on electricity companies' decisions to develop or operate wind farms.

One of the first environmental groups to interact with an electric utility was Western Resource Advocates (WRA), an environmental group formerly known as the Land and Water Fund of the Rockies. The restructuring and partial deregulation of the utilities, which started in the early 1990s, provided incentives for environmental groups such as WRA and utilities such as Xcel Energy to work together. The relationship between WRA and Xcel was initially contentious; although WRA expended many resources trying to force Xcel to invest in wind power, in early 1996 the utility withdrew from a planned wind project, and "the result of several years of adversarial jostling was essentially nothing."[93] In the same year, WRA and other environmental groups sought, but failed to obtain, a regulatory mandate from the Colorado Public Utilities Commission that would have required Xcel Energy (known at that time as the Public Service Company of Colorado) to add over 10 MW of renewable energy to its capacity.

Because they could not obtain a regulatory mandate, some environmental groups—including WRA—moved away from a "strictly adversarial" role and eventually agreed to Xcel's proposal to create a green-pricing program.[94] Under this program, called Windsource, customers choose to pay more on their electricity bills to purchase electricity from a new wind power project. According to one former WRA organizer,

> In the early '90s, when gas prices were low, everyone was invested in these new natural gas plants. We argued that the company should be diversifying into renewable energy, so there was some contention about the level of renewable energy the company was investing in through their resource planning process. We were disappointed that Xcel pulled out of a wind project in Wyoming. We thought the best next step forward was to get them to do this [Windsource] voluntary program. We reached a settlement with them in which they would bring forward this program.[95]

Another organizer emphasizes that WRA supported the program because it was an opportunity to develop new wind farms: "Xcel agreed that they would find customers and sign them up, and if they got enough customers, they would build a wind farm to supply the customers who had signed up. They also made a commitment to continue to increase the number of wind farms on their system as more and more customers signed up."[96]

However, WRA went beyond simply approving Xcel's plans to launch a green-pricing program; it also formed a partnership for community-based marketing. The partnership between WRA and Xcel sought to "lend credibility to the product and the marketing message and use grassroots organizing techniques to reach a broader set of potential customers cost-effectively."[97] Initially, there were a number of disagreements over Windsource between environmental groups and Xcel. Some environmental groups—including the Sierra Club and the Audubon Society—had concerns about the impacts on birds. Xcel and WRA were in disagreement over the pricing of the program: while WRA favored a price that reflected the actual cost of production, Xcel wanted to set the price based on willingness to pay. More importantly, WRA wanted to use a marketing strategy that highlighted the negative impacts of Xcel's coal power plants—a strategy vehemently opposed by Xcel, which has ten fossil-fuel power plants in Colorado alone. However, the initial disagreements were overcome after Xcel conducted a National Environmental Policy Act review, concessions were made on both sides regarding pricing, and WRA decided to use a marketing strategy that, in the words of a popular song, "accentuates the positive."[98]

Xcel started marketing the Windsource program in early 1997. Initially, Xcel planned to build 10 MW of wind power capacity, but sign-ups for the program exceeded expectations and Xcel built more wind farms to meet growing demand. Because WRA and other environmental groups developed a grassroots, community-based marketing campaign that collaborated with cities, schools, small and large businesses, and the general public, the Windsource program was a success. By the end of 2001, over 60 MW of wind capacity had been built in Colorado due to Windsource demand; by the end of 2004, the program was the largest green-pricing program in the country, with over forty thousand participants.[99] And by 2008, Windsource had consolidated this position, with over seventy thousand customers. In addition to individuals, the Windsource customers included numerous companies (IBM, Rocky Mountain Steel Mills, the Coors Brewing Company) and cities (Boulder and Denver).

The expansion of the Windsource program and Xcel's growing commitment to investing in wind farms were, to a large degree, results of environmental organizations' actions. By alternating between confrontational and cooperative strategies, environmental organizations contributed to Xcel's decision to develop and operate wind farms. As one Xcel manager recognizes, in the beginning the program was simply a reaction to environmental groups' pressure: "At that point, it was more of a reactive position that was undertaken—if this is the right word—to appease, or to satisfy the environmental groups here in Colorado. Frankly, at that point, I don't think the environmental aspect of it was fundamental to our CEO's decision."[100]

One WRA organizer described the role of the environmental community this way:

I don't think that this program would have happened without the involvement of the advocacy community. It wasn't just Western Resource Advocates; the Colorado Renewable Energy Society was also very engaged with this; the Sierra Club helped spread the word on the program. There were a number of groups that were actively trying to push Xcel. At the time, we really didn't have any utility-scale renewable energy in the state. The first 5 MW and then 10 MW were through the Windsource program. The other benefit of this program is that it really did open the door for a lot more renewable energy development down the line through other programs, including the renewable energy standard that was passed here.[101]

Another WRA organizer emphasizes that the Windsource program is number one in the country, despite the fact that Xcel spent less than other utilities on marketing, because of environmental groups' strong involvement:

I think the grassroots marketing and the one-on-one outreach that the environmental community did definitely helped build a strong base for support. In terms of Xcel's marketing, relative to the marketing that other utilities have done, it has been fairly simple. The majority of the marketing that they have done has been through utility bills and brochures that they mail out to customers. Other utilities have used a lot of innovative marketing strategies, but I wouldn't say Xcel tapped into those as much as others did. So, I think a big part of the program's success was the widespread grassroots support they had from environmental advocates.[102]

Environmental groups did not stop at marketing the Windsource program; they also pressured Xcel to add more wind and other forms of renewable energy to its portfolio. For example, in 2003 Xcel argued that in order to meet future demands for electricity, it would need to build a new 750 MW coal power plant. This plan was met with opposition from WRA and other environmental nonprofits, which argued in front of the Colorado Public Utilities Commission that the new coal power plant ignored the carbon dioxide regulatory risk. In the end, however, Xcel and the environmental groups agreed to compromise: while Xcel would build the new coal power plant, it would also meet a number of the environmental community's demands. Most notably, under pressure from environmental groups Xcel agreed to install advanced pollution-control equipment for two older coal power plants and to add up to 500 MW of wind power.[103]

Environmental groups also pressured Xcel indirectly, through the Colorado Public Utilities Commission. In accordance with the commission's rules, Xcel published a request for proposals for new electricity supply in 1999. Enron Wind submitted a proposal for a 162 MW wind farm in Lamar, Colorado; however, Xcel concluded that an "all natural gas" portfolio would satisfy the requirements of the least-cost solution. Xcel assumed that the cost of natural gas would stay low and that the ancillary costs of wind farms (which come from factors such as the need to have other power plants ready to generate electricity in case the wind doesn't blow) were high. Environmental advocates argued that the cost of gas would go higher and that the ancillary costs of wind farms were lower than Xcel's estimate. Somewhat surprisingly, the Colorado Public Utilities Commission sided with the environmental advocates and instructed Xcel to buy the output of the Lamar wind farm, making it clear that its decision was based not on environmental externalities but on the fact that wind would likely lower the cost of electricity.[104]

Another indirect pressure on Xcel came through a renewable portfolio standard (RPS). In 2002, an RPS that asked utilities to obtain 10 percent of their electricity from renewable sources by 2010 was introduced in the Colorado legislature by environmental advocates. Groups including WRA testified in favor of the bill, while Xcel testified against the bill, arguing that it would increase consumers' electricity bills. The bill was defeated in 2002, as well as in 2003 and 2004. Sensing that the bill would not have a chance to pass in the legislature, a number of local environmental groups mobilized to introduce a ballot initiative—known as Amendment 37—that would require the state's largest utilities to obtain 10 percent of their electricity from renewable energy by 2015. The environmental groups managed to get the sixty thousand signatures to get this amendment on the ballot relatively easily; however, their public campaign to convince voters to approve it met significant opposition.

Environmental groups from Colorado, in coalition with unions, economic development councils, and even cities, campaigned under the slogan "Cleaner Air, Cheaper Energy," arguing that the RPS would not only help the environment but also provide jobs and save money. Xcel Energy, the Colorado Rural Electric Association, and other groups organized a lobby group with the slogans "Right Idea, Wrong Solution" and "Say No to Unfunded Mandates." They argued that Amendment 37 would increase the cost of electricity for residential consumers, and that consumers already had the choice of purchasing wind power through the Windsource program. Ultimately, Amendment 37 passed by 54 percent to 46 percent and Colorado adopted an RPS that required Xcel and other utilities to obtain 10 percent of their electricity from renewable sources by 2015.[105]

While the above cases illustrate the role of Colorado-based environmental groups in Xcel's transformation, it is important to note that groups and activists in other states have also played significant roles in the decisions of electric utilities. In Minnesota, for example, the Izaak Walton League and other environmental groups have been wind power advocates since the early 1990s. Minnesota environmental groups were instrumental in getting the state to order Northern States Power—an Xcel subsidiary—to build 400 MW of wind plants in 1999. In fact, the 1999 boom in the wind energy industry in Minnesota was credited to "a small band of environmentalists"; as one American Wind Energy Association representative argued, this boom was "the culmination of decades of effort by renewable energy advocates in the Upper Midwest."[106] And in 2007, Minnesota environmental groups had a decisive influence on the governor's decision to adopt the strongest RPS in the country, which requires that electric utilities obtain one-quarter of their energy from renewable resources by 2025.

In Texas, Environmental Defense (ED) and the Natural Resources Defense Council (NRDC) organized a high-publicity campaign against TXU—the largest electric utility in Texas. In 2005, having considerable influence in the Texas political establishment, TXU was able to obtain from the governor an executive order to fast-track the permit process for eleven new coal-burning power plants. When Fred Krupp, president of ED, wrote to TXU's chairman and asked for a meeting, he was informed that TXU was on a fast track to build the plants and had the governor of Texas on its side.[107] Governor Perry incensed environmental activists even more when, echoing the nuclear lobby's criticism of antinuclear activists in Denmark more than two decades before, he argued that opponents to the TXU plan simply wanted to "return Texas to the era of the horse and buggy."[108]

The environmental community responded by mobilizing on multiple fronts. First, ED, the NRDC, and other environmental groups sued the governor, arguing that the fast-track order exceeded his authority. Then, they created a website—Stoptxu.com—on which they posted information about TXU's plans. Next, ED sued the Texas Commission on Environmental Quality, trying to stop permits for the coal plants. It also launched an "ad blitz" campaign and published articles in Texas newspapers and regional editions of the *Wall Street Journal.* One of the ads read "Good news: Some power companies are decreasing global warming pollution. Bad news: Yours is about to double it."[109] Additionally, ED publicized a fact sheet with the title "More Power than We Need, More Pollution than We Can Afford," arguing—among other things—that the eleven new plants would more than double TXU's emissions of carbon dioxide from 55 million tons a year to 133 million tons a year.[110] At the same time, more-radical groups such as Rainforest Action Network staged "die in" protests outside the TXU financiers' offices.[111]

A number of environmental groups—ED, the NRDC, Public Citizen, the Sierra Club—also created a broad coalition that included farmers, religious groups, and a number of Texas mayors concerned about local air pollution. An organizer from ED recollects how this happened:

We had local land owners and environmental groups, but we also had mayors in big cities who were worrying about their own clean air. We had a group called Texas Business for Clean Air, who were folks who were worried that new emissions from these plants would be more difficult for them to expand their own businesses or to attract workers who don't want dirty air when they go out to run or take their kids to the soccer games. We also had evangelicals; the Texas Baptists came out against these plants. So, it was a very broad, diverse group. There really was an outpouring of support. We got some emergency help from national foundations who gave money outside of their normal board cycle to help us. I literally had three people who gave me checks, unsolicited, of $10,000–$50,000. I have been doing this for nineteen years and nobody gives you money unsolicited, at least not big numbers. Three people had heard about the fight, they were concerned, they had talked to me. They put a check in my pocket. I thought: great! I put it in my pocket to look at it later, and it was five figures! That is unique... It became a big deal because it was about our kids' health in Texas and it was about global warming, and people rallied around it.[112]

Environmental groups had an unexpected opportunity in early 2007, when two private equity firms—Kohlberg Kravis Roberts & Company and Texas Pacific Group—offered to buy TXU in what was considered at the time the largest leveraged buyout ever attempted. Concerned about costly litigation, the buyers contacted ED and basically asked "what would it take" to gain their support.[113] Environmental Defense sent out a negotiator who reached a preliminary agreement with the two private equity firms; as he recollects:

On short notice I flew to San Francisco and spent seventeen hours negotiating and looking at things that they would agree to do that would make a difference with regard to their [greenhouse] emissions and also what they would do in regard to other activities like lobbying for federal legislation on global warming and a number of other things. There was a pretty good list, and it got an amazing amount of publicity; on the front page of the *New York Times*, on the radio in places like Australia, Japan, Europe, and other international press. This was a model for the kind of things that investors ought to be concerned about with regard to their investments; it showed that here is going to be a new world that will have limits on carbon dioxide emissions, and that if you don't plan for that when you are merging or acquiring a company, you may be set up with big future obligations. This was an acquisition, and folks in the acquisition world understanding that the economy is going to be changing significantly in the next few years. If you don't understand that, you are at grave risk.[114]

In a move that was described in the media as presaging a heightened role for environmental activists in mergers and acquisitions, in March 2007 ED hired an investment bank to advise it as it continued to negotiate the fine print with the private equity firms. According to Fred Krupp, the president of ED, the organization hired the bankers because it had never been involved in a buyout, and it wanted to make sure that it had the best expertise available. He also adds that ED is pleased with the two private equity firms because they made a set of commitments that environmentalists can feel good about. Some environmentalists were less pleased with the deal; some people involved in the deal argued that TXU had planned to reduce the number of coal-fired power plants to five or six, from eleven, anyway.[115] Nevertheless, most individuals inside as well as outside the environmental community recognized that ED and other environmental organizations achieved a significant victory because they influenced TXU's decision to build fewer power plants, to significantly increase its energy efficiency programs, and to double its purchase of wind power. Indeed, an important consequence of the environmental activists' efforts was that TXU became in early 2009 one of the utilities that offers Green-e Energy certified wind power to its commercial customers.[116] As *New York Times* columnist Thomas Friedman noted, the environmental campaign against TXU showed to the world that "truth plus passion plus the Internet can create an irresistible tide for change."[117]

National and regional organizations that specialize in litigation and lobbying—ED, the NRDC, and WRA—are not the only environmental groups that influenced electricity companies' decisions to invest in wind farms. Large and small environmental groups that specialize in direct action have also played important roles. For example, Rising Tide and Earth First! staged vigorous protests against Duke Energy's plan to build a coal power plant in Cliffside, North Carolina. In November 2007, a number of climate change activists protested at Duke Energy's headquarters in Charlotte, North Carolina, and two were arrested.[118] As part of the Fossil Fools International Day of Action, on April 1, 2008, several environmental activists locked themselves to bulldozers while others roped off the plant site with "Global Warming Crime Scene" tape; the protests resulted in eight arrests.[119] Other organizations soon joined in the fight against Duke's Cliffside coal plant. In April 2009, a coalition of large national organizations (including Greenpeace USA); medium-size regional organizations (the Southern Alliance for Clean Energy, for example); and small, local organizations (such as Upper Watauga Riverkeeper) gathered about three hundred people to protest in front of Duke's headquarters.[120] The protests resulted in the arrest of forty-four activists.[121]

Feeling the heat from the environmental protests and anticipating even more protests against new coal power plants in the near future, Duke Energy began investing in wind power. In 2007, it acquired Tierra Energy, a company based in Texas, which had 180 MW of wind assets almost completed. In 2008, Duke purchased Catamount Energy, a company based in Vermont, adding another 300 MW of wind energy to its portfolio. By the end of 2008, Duke had almost 500 MW of wind power in operation; it also planned to develop an additional 5 GW in the next few years.

Duke's decision to invest in wind power was influenced not only by environmental groups' direct actions; it was also influenced by a perceived change in the social context surrounding the climate change issue. Sensing that climate change regulation was imminent, in 2001 Duke Energy CEO James Rogers told a meeting of fellow electricity company CEOs that they should work to pass a federal carbon cap. At that time, the electricity sector considered this a heresy and Rogers paid the "ultimate price" for a CEO: he was not invited to play golf with fellow energy executives any more![122] Rogers became known as an advocate of cap-and-trade regulation, and in

2007, Duke Energy became one of the founding partners of the U.S. Climate Action Partnership (USCAP), an industry group that called on President George W. Bush "to fight global warming by limiting greenhouse gases, funding research into renewable energy and creating a market for carbon dioxide emissions."[123] Moreover, in the same year Rogers presented a plan to "decarbonize" Duke Energy by 2050 and introduced the Save-a-Watt plan, which charges Duke's customers in a program not to build new power plants rather than to do so. As Rogers explained his decisions, "I wanted to get out ahead of it. It's the old saw—'If you're not at the table, you're going to be on the menu.'"[124]

Anticipating climate change regulation, Duke senior managers decided to increase investments in wind power and to shape the policymaking process to their advantage. One manager describes these decisions as the result of the desire to gain a competitive advantage:

> We have seen a growing public demand for clean, renewable energy. Companies that have the foresight to meet that demand are going to be better positioned to meet that demand as we progress in the twenty-first century.... We have seen this [climate change regulation] coming for quite some time. I think the new Obama administration is a clear sign that this is going to be an agenda item in Washington in 2009 and 2010. Our CEO Jim Rogers is the undisputed thought leader in the energy industry on carbon cap and trade. While some energy companies' CEOs would rather wait on the sidelines and take a "wait and see" approach, he is very actively involved in shaping policy.[125]

But despite all his talk about investing in wind farms to "build bridges to a low carbon future," and despite the fact that Rogers collaborated with environmental organizations to promote a cap-and-trade system, many environmental activists think that Duke's CEO is not doing enough. One reason is that, as the head of the environmental group Clean Air Watch notes, "This [cap-and-trade] bill gives huge windfall profits to a company that buys a lot of coal, like Duke. I happen to think that it's immoral. In a sense, you're paying the polluter. You're rewarding the very companies that are the source of the problem."[126]

Another reason for discontent is that Rogers supports building new nuclear power plants. According to a Duke Energy manager,

> One thing Jim Rogers has said repeatedly—and we all believe this steadfastly—is if you are serious about climate change, you have to be serious about nuclear energy as well. It is virtually emissions free, with no greenhouse gases. Obviously, it has other issues associated with it, but the cost to build nuclear is far greater than two decades ago, so we haven't had any new nuclear come online in the U.S. for the past decade and a half. There was almost a moratorium on building new nuclear. But this informal moratorium seems to be dissolving as there have been over one hundred applications for new nuclear power

sites made to the nuclear power commission. So we are exploring our options to build new nuclear power plants.[127]

The importance of gaining a competitive advantage in a future carbon-regulated world has also been an important factor in other utilities' decisions to invest in wind power. For example, NextEra Energy (which was known as FPL Energy before January 2009) started investing in wind in 1989 because it was looking for opportunities to make strategic investments; in the words of one NextEra manager, "It was a passive investment in a distressed asset. We were looking for opportunities to make some strategic investments." By 2000, however, NextEra managers realized that they could gain a competitive advantage because there was a growing demand for clean energy, and they already had some experience in developing and operating wind farms.

> The real growth in our wind business really occurred in 2000 and beyond. This was a time period when there was significant growth in natural gas generation; a lot of competition in the natural gas generation business. We did some of that as well, but our president and his leadership team looked at the landscape and felt that we have or could have a competitive advantage in the wind business because we had lots of years of operating experience.... At the end of the day, this is a comment that you will hear from our executives a lot: the wind business is good for our shareholders, for our customers, and for the environment.[128]

Environmental groups and activists have also influenced electricity companies' decisions to develop and operate wind farms by gradually changing their employees' values. For example, NextEra Energy managers recognize that energy-sector professionals are more interested in environmental protection today than they were a few decades ago. According to one NextEra Energy manager,

> Environmental responsibility and doing the right thing is at the core of what we are all about. I personally think that younger people are attracted to working for the largest renewable company in North America—but economics are obviously important. I think there is a greater awareness about the environment now than when I entered the workforce, about twenty years ago. We hear from job candidates that they want to work for a renewable energy company. One of the striking things is that, when you go out to any of our wind farms, you will always see tremendous, tremendous amount of pride in the workforce in what they are doing. I've worked at two other energy companies, but only at NextEra I've seen this incredible amount of pride, the feeling that we are doing the right thing and we are making a difference.[129]

Other companies that develop and operate wind farms recognize the importance of having a green corporate culture. But, as one Xcel Energy manager points out, the corporate culture's shade of green is a reflection of the strength of the environmental community:

Our corporate culture has really changed over the last decade. Originally, I think that there was moderate support for the voluntary Windsource program. But after the previous chairman retired, we have Dick Kelly, who is now the chairman and CEO of Xcel Energy, and he has shown himself to be quite an environmental supporter, in terms of recognizing the importance of the climate change issue, and really changing the corporation toward a recognition of environmental leadership much more than the previous chairman.... The other thing to remember is that utilities are creatures of the state. Regulation of utilities is local. If you look at a state like Minnesota, for instance, it has a very long history of an environmental ethic in that state, and it goes into resource planning, energy conservation, and efficiency, and it shows up in support for renewables. In Colorado, the history of the environmental movement is not as long as in Minnesota, but it has come along. So the point is, we have a leader that recognizes these things, and we have citizens throughout our states that are supportive of these things.[130]

James Dehlsen, who has interacted with electricity companies since the early 1980s, also sees an important generational change that has impacted their corporate cultures:

In the 1980s, utilities were still pretty much of the mindset that big, central power plants were the solution. The bigger the better: big nuclear plants, big coal plants, that's how real men make electricity! But at some point, there was a change in that mindset. I think the reason that eventually changed was because that old guard went through a generational change where the guys that started coming in the 1980s and the 1990s were a younger group of managers that did have an understanding of the environmental priorities they should be paying attention to. In the '60s, '70s, and '80s, there was a huge expansion in nuclear and big central coal-powered plants. These were gigantic projects with great volume and huge amounts of financing. So, you had a whole level of utility management that was keen to that. But then, as those days started passing, you had new managers coming in and they had their kids telling them they need to be more environmentally sensitive.[131]

Environmental activists contributed to the greening of the electricity sector in the United States not only by engaging in protests, litigation, lobbying, and marketing, but also by founding companies that specialize in wind-farm development. For example, Eric Blank has worked for the environmental nonprofit WRA based in Boulder, Colorado. During the mid-1990s, Blank helped pioneer the field of wind energy marketing, working in conjunction with several Colorado utilities. In 1999, realizing the potential for marketing and developing wind energy in newly deregulated states, Blank cofounded Community Energy together with Brent Alderfer, a lawyer working for the Colorado Public Utilities Commission. The company started by building two turbines funded by grants from two environmental groups, the Clean Air

Council and WRA. But Community Energy was "a team on a mission—a mission to supply the demand for fuel-free energy through the construction of new wind farms."[132] By most measures, its mission has been accomplished: by 2004 it had more than forty thousand residential customers in seven states, and five hundred business and institutional customers, becoming the largest retailer of wind power in the United States.[133] In 2006, it received the Green Power Pioneer Award from the U.S. Environmental Protection Agency and Department of Energy, and it had found—in the words of its president—"the perfect marriage," with Iberdrola, the largest renewable energy operator in the world.[134]

Finally, it is important to note that environmental activists and organizations also helped developers overcome local opposition to wind farms. Perhaps the best known case is that of Cape Wind, the first offshore wind farm in the United States. In 2001, Jim Gordon, the president of Energy Management Inc. (EMI), announced that his company intended to build an offshore wind farm on Horseshoe Shoal in Nantucket Sound, Massachusetts. Within weeks of the announcement, a number of local residents who owned expensive property facing Nantucket Sound formed the Alliance to Protect Nantucket Sound (APNS). The APNS immediately launched probably the most well-funded Not-In-My-Backyard (or, more accurately, Not-In-My-Frontyard) public-relations campaign ever organized in the United States.[135] Its goal was to stop the Cape Wind project by dragging out the regulatory process. To this end, members of APNS launched television and newspaper ads, collected signatures, staged protests, filed lawsuits, and lobbied heavily.[136]

This campaign delayed the Cape Wind project for many years. Although the review process was initially expected to last until 2004, Cape Wind was only approved by the U.S. Interior Secretary Ken Salazar in April 2010. Given how well-funded and professionally organized the APNS campaign was, it is surprising that the project was eventually approved and that it will probably begin to be built almost a decade after it was proposed. The APNS and its influential political allies—which included, among others, Massachusetts Senator Edward Kennedy and Massachusetts Governor Williard Mitt Romney—did not succeed in killing the Cape Wind project partly because of Jim Gordon and his staff's tenacity. EMI's president and staff fought tooth and nail to keep the project alive, not only because it made economic sense but also because they believed it was the right thing to do for the environment. According to Marc Rodgers, Communications Director for EMI, "There is a real environmental sensibility among people who work at the company. In the case of Jim Gordon, our company president, [environmentalism] is a real passion. [...] I think that there is one thing that our company can be credited for, in addition to the strength of the proposal, and that's perseverance. I think that many companies would not have chosen to stay the course over as many years as we have."[137]

Even with EMI's perseverance, however, it is unlikely that its offshore wind project would have survived APNS's attacks without the support it received from environmental groups. Marc Rodgers argues that virtually all major environmental groups were behind the Cape Wind project: "It is clear that all major environmental organizations of this country that work in the field of energy are pretty much all lined up in support of this project. For a lot of environmental organizations it was obvious that this was an example of exactly the kind of initiative and project they've been calling for for so long. And their support has become more substantive over the years, as they've seen that the review process has been thorough, that things are being looked at closely, and that there really are no red flags jumping out."[138]

The list of environmental groups that support the Cape Wind project includes large national organizations such as Greenpeace, the Natural Resources Defense Council, Friends of the Earth, and the World Wildlife Fund, as well as many regional and local organizations such as Massachusetts Climate Action Network, the Environmental League of Massachusetts, Cape & Islands Self Reliance, and Clean Power Now.[139] Environmental groups have frequently demonstrated in support of the Cape Wind project and used creative tactics to counteract APNS's campaign. For example, supporters of Clean Power Now, a grassroots organization founded on Cape Cod in 2003, staged a colorful protest at one of the U.S. Army Corps public hearings to discuss Cape Wind's draft Environmental Impact Statement. The demonstration attracted a lot of media attention because the protesters dressed in old-fashioned yachting costumes, carried ironic signs that read "Global Warming: A Longer Yachting Season", and shouted humorous slogans such as "Save our Sound! Save our Sound! Especially the View from My Compound!" or "Cape Wind makes our Blue Blood Boil! Let's get our power from Middle East Oil!"

Environmentalists also used creative tactics to pressure Senator Edward Kennedy to stop opposing Cape Wind. For example, Clean Power Now members protested and formed a "human windmill" under the senator's Boston office windows. Additionally, at a signing event for Kennedy's book *America Back on Track*, Greenpeace activists handed out fliers for a fake Kennedy book called *How I Killed America's First Offshore Wind Farm*. Due to environmental groups' strategy, the public and the mass media became increasingly supportive of the Cape Wind project. A 2004 Boston Globe article reflected public sentiment in its title "Cape Wind: Too Ugly for the Rich?"[140]

Conclusion

This chapter has shown that the electricity sector has "gone with the wind" —more specifically, it has undergone two transformations since the 1990s. First, small, traditional wind turbine manufacturers have become industrial heavyweights, and traditional power plant manufacturers have recognized that wind turbine manufacturing is big business. Second, new companies specializing in wind-farm development and operation have emerged, while growing numbers of electric utilities have invested in wind farms. This chapter has demonstrated that the environmental movement contributed to these transformations when environmental activists gained control of energy companies and professional societies, criticized the traditional logic of energy production, and offered practical solutions. By becoming entrepreneurs, innovators, advocates, and champions, environmental movement activists and sympathizers contributed to wind turbine manufacturing. By forming wind turbine cooperatives, founding wind-farm development companies, and constantly pressuring utility companies to invest in renewable energy, environmental organizations and activists also had a major impact on wind-farm development. Consequently, while in the past manufacturers of power plants and producers of electricity were interested only in producing the most reliable equipment or the cheapest possible electricity, they are now increasingly concerned about producing green equipment and electricity.

The case studies illustrate how environmental movement activists and organizations in Europe and North America have contributed to the gradual greening of the electricity sector. In Denmark, some of the activists involved in the Organization for Renewable Energy during the 1970s have become important entrepreneurs and inventors, while others have become wind power advocates. The open exchange of information—encouraged by organizations such as the Organization for Renewable Energy and the Nordic Folkecenter for Renewable Energy—has contributed to a diffusion of innovation in wind turbine technology and has benefited wind turbine manufacturers not only in Denmark but also in Germany and the United States. Additionally, the environmental movement has played a role in the emergence of wind power entrepreneurs, advocates, and champions in Germany and the United States.

Environmental movement activists and sympathizers have also contributed to the emergence of wind turbine cooperatives and independent power producers that specialize in wind-farm development in Europe and North America. And environmental organizations and activists have used both confrontational and cooperative tactics to pressure utilities to invest in wind farms, particularly in countries with deregulated electricity markets such as the United States. Therefore, although the electricity sector in the majority of countries remains predominantly "brown" and relies heavily on burning fossil fuels, green sprouts have emerged in some places partly because of a slow (but not silent) environmentalist spring. The concluding chapter will summarize the most important purposive consequences of the environmental movement's involvement, and will also briefly discuss some of its unintended consequences for the wind energy industry.

Conclusion

The Answer May Be Blowing in the Wind

Carefully calculating and taking into account some insecurity factors, wind energy will be able to contribute in the year 2020 at least 12 percent of global electricity consumption. . . . by the year 2025. . . . All renewable energies together would exceed 50 percent of the global electricity supply.
—World Wind Energy Association, "World Wind Energy Report 2008," http://www.wwindea.org/home/index.php?option=com_content&task= view&id=226&Itemid=43 (accessed May 2010)

The renewable share of world electricity generation [will] fall slightly, from 18 percent in 2005 to 15 percent in 2030, as growth in the consumption of both coal and natural gas in the electricity generation sector worldwide exceeds the growth in renewable sources of generation.
—U.S. Energy Information Administration, "International Energy Outlook 2008," http://www.eia.doe.gov/oiaf/ieo/electricity.html (accessed May 2009)

A Summary of the Environmental Winds of Change

Although approximately 1.6 billion people—a quarter of humanity—live without electricity today, for most people in the developed world life without electricity is inconceivable.[1] We take for granted the invisible force of electricity that brings life to our mobile phones, televisions, computers, refrigerators, and a myriad of other appliances, but we are generally ignorant about its origin. A survey of Americans conducted in 2008, for example, found that when asked where their electricity comes from, 35 percent of people said they do not know and about 23 percent said their electricity comes from "'electricity' or the 'electric company.' "[2] In fact, only 16 percent of respondents named coal as fuel for their electricity, and 7 percent named nuclear power, although coal and nuclear power account for about 50 percent and 19 percent respectively of the electricity used in U.S. homes. When it comes to our knowledge about electricity's origin, many of us are no different from children who think that milk comes from "the supermarket." This is troubling: as Robert Pogue Harrison (2002, 359) notes, "When the 'from' of the things we consume becomes not only remote but essentially unreal, the world we live in draws a veil over the earth we live on."

Global demand for electricity will continue to grow. According to the U.S. Energy Information Administration (EIA), "Over the next 25 years, the world will become increasingly dependent on electricity to meet its energy needs. Electricity is expected to remain the fastest-growing form of end-use energy worldwide through 2030, as it has been over the past several decades."[3] In fact, the EIA estimates that world net electricity generation will almost double from 17.3 trillion kWh in 2005 to 33.3 trillion kWh in 2030.

Experts' predictions about the future role of wind and other renewable sources of energy diverge considerably. Some scenarios are optimistic and predict that wind alone could produce almost 12 percent of electricity by 2020, and all renewable energies together could exceed 50 percent of the global electricity supply by 2030.[4] Other forecasts are less positive and predict that renewable energy will produce only 15 percent of electricity by 2030.[5]

Attempting to predict exactly how much wind and other renewable sources will contribute to electricity generation two decades from now is, obviously, futile. It is important to note, however, that even conservative estimates about the future of renewable energy predict significant growth for the near future. Many energy analysts agree that wind energy is already a mature industry and that interest in renewable sources of energy is rapidly growing worldwide. A significant milestone was reached in 2008, when added power capacity from wind and other renewable sources of energy in the European Union and the United States represented more than 50 percent of total added capacity—in other words, it exceeded added power capacity from gas, coal, oil, and nuclear power taken together. Another indicator of the rising importance of renewable energy is the fact that global investment in renewables has dramatically increased over the last few years, from $21 billion in 2004 to $120 billion in 2008.[6]

This book started from the observation that while wind power stands out as a renewable energy success story in some countries and regions, it has failed to reach its true potential in many countries and has had an uneven global development. The study offered an interpretation that differs from the two dominant approaches: the technological approach, which presents the development of the wind energy industry as the result of decreasing costs in wind power generation due to continuous improvements in technology and explains its uneven growth as the result of differences in technological styles; and the economic approach, which argues that variation in the growth of this industry results from disparities in economic forces such as energy policies and the deregulation of electric utilities. The book used a contentious politics or social movements perspective, which centers on the role played by social movements in industry emergence and development. It has shown that it is not possible to fully understand the specific development of the wind energy industry without examining the environmental movement's influence on the energy sector.

The book developed a model that centers on social and political factors, in particular on the role of environmental activists and organizations. The model built on social movement and industry creation theory and argued that the development of the wind energy industry is influenced by interactions between the environmental movement, the social context, and natural resources. The model predicted that the wind energy industry would grow faster in countries and regions that not only have good wind potential but also a favorable social context and an environmental movement that mobilizes in support of clean energy. The first chapter used quantitative analysis to test this prediction; results showed that the wind energy industry grows faster in countries that have strong environmental organizations, as well as high-quality wind and political allies who work with environmental organizations. Results also showed that countries that adopted renewable energy feed-in tariffs (FITS) are more likely to have a developed wind energy industry than other countries.

The book identified three main pathways through which the environmental movement influences the development of the wind energy industry. The first pathway is the influence that environmental activists and organizations have on energy policymakers' decisions to adopt and implement pro–renewable energy policies. Chapter 2 showed that environmental groups and activists have contributed to the adoption and implementation of renewable energy policies such as FITS in Germany, Denmark, and Spain. It showed that the origin of FITS can be traced to the emergence of grassroots groups and research institutes dedicated to promoting the use of renewable energy during the 1970s and 1980s. During the 1990s, as global climate change became the dominant issue on environmental groups' agendas, environmentalists advocated for the adoption of FITS and other pro–renewable energy policies and mobilized to defend their implementation when these policies were threatened by the fossil-fuel, nuclear, or utility lobbies. The success of environmentalists resulted from their ability to build large pro–renewable energy coalitions, as well as from the fact that they had allies among political elites. A social context characterized by an actively involved and unbiased mass media and favorable public opinion also contributed to their success.

Chapter 3 examined the way in which environmental groups and activists shaped the energy policymaking processes in countries such as the United Kingdom, the United States, and Canada, which have had somewhat less-vigorous environmentalist mobilizations and a less-favorable social context. This chapter showed that environmentalists' abilities to reach their goals are severely limited when they lack influential political allies and when they face a less-committed or biased mass media and less-favorable public opinion. Consequently, much of the progress on the renewable energy front in the United States and Canada has been made at the state and local level, where the social context is more favorable.

The second pathway is the influence that environmental groups and activists have on energy consumers. As chapter 4 demonstrated, the environmental movement played an essential role in creating consumer demand for clean energy. The chapter showed that U.S. environmental organizations influenced college, university, and corporation decisions to purchase renewable energy certificates (RECs) from wind power. The analysis demonstrated that, although environmental groups and activists can pressure organizations to change from outside through boycotts, protests, and shareholder activism, their most significant impact is to create change from inside. The environmental movement changed the ideological commitment of organizational members, turning them into environmental mediators. Environmental organizations pushed colleges and universities for green-power purchases by organizing student campaigns for clean energy and by coordinating a network of top-level administrators who were committed to addressing climate change. Environmental groups also contributed to corporations' decisions to purchase green power mainly by offering resources to mid-level employees and environmental managers.

The third pathway is the influence of the environmental movement on energy professionals. Chapter 5 showed that environmental activists and organizations changed the electricity sector's rationale. Manufacturers of power plants and producers of electricity are no longer interested only in producing the most reliable equipment or the cheapest possible electricity— they are increasingly interested in producing the greenest possible electricity. This chapter demonstrated that the environmental movement contributed to the greening of the electricity sector in two main ways. First, environmental movement activists and sympathizers made an essential contribution to wind turbine manufacturing by becoming entrepreneurs, innovators, advocates, and champions. Second, environmental organizations and activists had a major impact on wind-farm development by forming wind turbine cooperatives, particularly in countries such as Denmark and Germany. They also contributed to wind-farm development by founding wind-farm companies, pressuring utility companies to invest in renewable energy, and aiding developers overcome opposition to wind farms, particularly in countries with deregulated electricity markets such as the United States and the United Kingdom.

The three main pathways identified in this heuristic model are simplified descriptions of the processes through which the environmental movement impacts the wind energy industry. In reality, as the empirical analyses have shown, these pathways overlap; for example, many environmental groups and activists are simultaneously lobbying for clean energy policies, educating electricity consumers about green power, and pressuring electric utilities to invest in renewable energy. Moreover, these pathways illustrate only the environmental movement's purposive consequences. But, as the model specifies, environmental groups' actions also have unintended consequences. For example, by pressuring national governments to join international negotiations on climate change and reduce global emissions of greenhouse gases, environmental groups such as those involved in the Climate Action Network contribute to the development of wind and other renewable energy projects in developing countries.[7]

Industrialized countries that ratified the Kyoto Protocol have the option to use the most cost-effective reduction in greenhouse gases. They use three mechanisms: clean development, joint implementation, and emissions trading. Many environmental groups have opposed the use of these mechanisms, arguing that they offer a way for developed countries to avoid much-needed domestic reductions of greenhouse gases. Yet, these mechanisms have resulted in technology transfer and significant development of wind energy projects in developing countries. For example, by the beginning of 2009 over 538 wind energy projects were in the Clean Development Mechanism (CDM) pipeline, totaling an installed capacity of 20.434 GW.[8] Most of these projects are installed in China and India. In China 90 percent of wind energy projects have applied for CDM registration, and 254 projects totaling 13 GW of capacity are in the pipeline. In India, 231 wind energy projects totaling more than 4 GW of capacity were in the pipeline by the beginning of 2009.[9]

It is also important to point out that although the model depicts the environmental movement in a single box, the movement is not a monolithic force. As the definition in this book's introduction emphasizes, the environmental movement is a loose network of interactions between formal organizations and informal groups engaged in collective actions and motivated by concern about environmental issues. The environmental movement is characterized by different subcultures and discourses, ranging from environmental problem-solving to green radicalism.[10] Not surprisingly, there are some differences among environmental organizations regarding their level of support for wind farms. Some organizations, which closely follow the "Small is beautiful" principle, favor small-scale or community wind farms. But the overwhelming majority of environmental groups supports all wind farms because the global climate change threat is so big. Consider the following statement by John Passacantando, executive director of Greenpeace USA, which summarizes many environmental organizations' response to NIMBY opposition to wind farms: "I respect people who wage NIMBY battles—the environmental movement was founded on people protecting their local, sacred areas. But today, solving the climate crisis has become so urgent that it trumps NIMBYism. It's as simple as that."[11]

It is a common misconception, however, that the environmental movement is divided over wind energy. The opposition to certain wind farms has often been portrayed in the mass media as a fight between different environmental factions.[12] In the case of Cape Wind, the first proposed offshore wind farm in

America, the widespread perception of a split in the environmental community is partly due to opposition from Robert Kennedy Jr., who is a senior attorney at the Natural Resources Defense Council (NRDC). However, his personal opinion does not represent the NRDC's position; as the organization declared in one of its statements of support for this wind farm, "Cape Wind has been under thorough review for seven years. NRDC participated in that process, and we concluded that the project's benefits will clearly outweigh its costs. At this point, we have to move forward."[13]

The resistance to Cape Wind and other wind projects comes most often from small but vocal groups that are not associated with the environmental movement. In fact, many of the groups opposed to the Cape Wind project have ties to the oil or coal industries.[14] Although project opponents and the mass media have claimed that the Audubon Society opposes the Cape Wind project, the organization conducted independent research that stated, "Our preliminary conclusion is that the project would not pose a threat to avian species."[15] The organization's official position on wind energy is the following: "Audubon strongly supports properly-sited wind power as a clean alternative energy source that reduces the threat of global warming. Wind power facilities should be planned, sited, and operated to minimize negative impacts on bird and wildlife populations."[16]

In the United Kingdom, wind-farm opponents frequently name the Royal Society for the Protection of Birds (RSPB) as their ally. However, this old and respected organization has reiterated its support for wind energy and stated that its opposition to wind farms is minimal. According to one of its reports, "Wind power is the most advanced renewable technology available at a large scale during this time period. For this reason, the RSPB supports a significant growth in offshore and onshore wind power generation in the U.K....We scrutinize hundreds of wind farm applications every year to determine their likely wildlife impacts, and object to about 7 percent, because they threaten bird populations."[17]

While this book focused on the environmental movement's influence on the global development of the wind energy industry, it also recognized the important role played by various technological factors. A major technological issue is that of "penetration," which refers to the amount of electricity produced from wind compared to the total generation capacity. The limit of wind power penetration depends on a number of factors, including the type of existing generating plants, capacity for storage, and demand management. Because most grids have reserve generating and transmission capacity to allow for equipment failures, this reserve capacity is often used to regulate the varying power generation of wind farms. A number of studies attempt to calculate maximum wind power penetration, or the amount of wind power that can be integrated into an existing grid without destabilizing it or significantly increasing the price of electricity. Some studies have calculated that maximum wind power penetration is 20 percent of total energy consumption, or the amount of wind power currently generated in Denmark.[18] Other studies find that in countries such as the United Kingdom it would be possible, using wind power, to accommodate 50 percent of total power delivered at modest cost increases.[19]

Developing technological solutions for problems specific to transmission and to offshore wind farms is essential for the industry's future growth. Wind power proponents argue that new high-voltage direct current (HVDC) power systems will have to be constructed in the near future because they are less expensive and suffer lower electrical losses over long distances. For example, a study that looked at European-wide adoption of renewable energy and interlinking power grids using HVDC cables suggested that the European Union's entire power usage could come from renewable sources, with 70 percent of total energy from wind.[20]

Future growth in the wind energy industry will come mostly from offshore wind farms. Offshore wind potential is very large because offshore wind is stronger and more constant. MacKay (2008, 62) calculates that using current technology, offshore winds in the United Kingdom could deliver a power of 48 kWh per day per person. According to his estimates, this is significantly more than the amount of energy used for lighting, heating, and cooling the entire country. But important technological challenges remain for offshore wind. Offshore wind turbines have to be much more resistant to corrosion than onshore wind turbines. They also have to be bigger and, sometimes, they have to be built at deep sea to produce electricity more efficiently. German manufacturers such as Multibrid and Repower already produce gigantic 5 MW wind turbines for offshore generation, while a Norwegian company named Sway has developed a deep-water system capable of supporting 5 MW wind turbines.[21] A few Northern European countries (Denmark, Sweden, Germany, and Norway) have started a research-and-development cooperation program called the Joint Declaration on Cooperation in the Field of Research on Offshore Wind Energy Deployment.

This book also acknowledged the role of economic factors such as job creation in wind turbine manufacturing, and additional income for farmers. For example, in 2008 a total of 36,249 jobs were directly created and an additional 48,051 jobs were indirectly created by the German wind energy industry. The wind energy industry is also a major employer in other European countries: in Spain and in Denmark the wind energy industry employs 37,730 and 23,500 people respectively.[22] In fact, as one study argues, "cities such as Nakskov and Esbjerg in Denmark, the region of Schleswig-Holstein in Germany, and the region of Navarre in Spain are all examples of areas where the wind energy sector continues to have a dramatic impact on the local economies and overall employment."[23] Additionally, wind farms bring economic benefits to farmers who lease part of their land to developers. For example, one author noted that in the United States, a farmer "who leases a quarter acre of cropland to the local utility as a site for a wind turbine can typically earn $2,000 a year in royalties from the electricity produced. In a good year, that same plot can produce $100 worth of corn."[24]

The role of economic factors is even greater in developing countries. In China, economic imperative has been a major driving force behind the recent growth of the wind energy industry. The Chinese government supports local manufacturing of wind turbines because wind farms benefit the local economy through employment in wind-farm construction and maintenance, as well as through grid extension for rural electrification. While in the past, imported wind turbines dominated the market, in recent years a growing market and clear policy direction have encouraged domestic production. As a result, by 2008 over forty Chinese manufacturers were involved in wind energy and two companies—Goldwind and Sinovel—were already among the top ten wind turbine manufacturers in the world.[25]

Job creation, however, is only one of the economic drivers for wind energy in developing countries; another one is the need to find alternatives to relying on a conventional power grid that can often be erratic. Somewhat paradoxically, since wind power is considered less reliable than conventional electricity generation, wind turbines provide relatively reliable electricity for industrial enterprises in countries that have underdeveloped and unreliable power distribution systems. This is particularly the case in India, where almost 70 percent of the demand for wind turbines comes from industrial users who are dissatisfied with the cost and reliability of electricity provided by state-owned electricity companies.[26]

Although it was not discussed at length in the book, another factor that has contributed to the growth of the wind energy industry is natural resources endowment. In addition to the availability of excellent wind, this industry is also dependent on the availability—or rather, the unavailability—of fossil fuels and other sources of energy. Spain, for example, depends upon imports for the bulk of its energy needs; some regions, such as Navarra, have no coal, oil, or gas fields, and no thermal, nuclear, or large hydroelectric power stations. Therefore, the Spanish national government and the regional government of Navarre have strongly encouraged the development of wind energy and other renewables. Not surprisingly, Navarre is today home to four wind turbine assembly and blade manufacturing factories, two component manufacturing factories, and the largest wind turbine testing laboratory in the world.[27] In contrast, Norway also has good wind potential but also abundant oil, natural gas, and hydropower resources. Because it produces almost 99 percent of its electricity from hydropower, Norway has only recently begun to encourage the development of wind power.[28]

Implications for Research on Industry Development

How does this research on the development of the wind energy industry inform future studies of industry emergence and growth? The book's most important contribution—from a theoretical standpoint—is to bring social movements into the study of market formation and industry growth. Other studies have also emphasized the significance of social movement activism and collective actions for both new and established industries. Rao (2009, 5), for example, argues that most research in economics or economic sociology "neglected to understand how the *joined hands* of activists and their recruits make or break radical innovation in markets." His research shows that "market rebels"—activists who challenge the status quo—can shape new or existing markets when they articulate "hot causes" that arouse emotions and create communities of members by relying on "cool mobilization" that signals the identity of community members and sustains their commitment. The importance of having an emotional, or hot, cause for campaigns to achieve "stickiness" and become successful is also emphasized by Soule (2009). Indeed, mobilizations around hot causes using creative tactics have played essential roles in the adoption of new technologies and practices in various industries, including the computer, automobile, brewery, biotechnology, chemical, and apparel industries (Smith, Sonnenfeld, and Pellow 2006; Rao 2009; Soule 2009).

But, while existing studies of the impact of activism on industry emergence or evolution have focused mostly on what activists do and how targeted companies react—in other words, on the dynamic between activists' mobilization and companies' responses—this book's research emphasized that social movements can affect market formation through four different pathways. One pathway is top-down, influencing the adoption and implementation of specific policies. Another is bottom-up, changing consumers' perceptions and creating demand for a new product. Yet another pathway is intermediate, changing the rationale specific to industrial sectors. A final one is unintended, influencing supranational agreements that shape the regional distribution of capital and technology.

The book also argued that these pathways of influence are shaped by interactions that take place within the social and natural contexts in which movements operate. More specifically, to be highly successful, movements have to not only mobilize around hot causes using cool tactics, but also be able to find political allies, have access to a committed and unbiased media, and align with the general public's attitudes. Additionally, regional and national natural resources may either impede or facilitate a social movement's ability to influence the policymaking process, create consumer demand for innovative products, or change the way professionals within a certain industrial sector think about their mission.

Consequently, this book suggests that future studies of market formation and evolution should analyze the role of social movements in detail only if certain conditions are met. First, there has to be a significant and sustained mobilization effort by social movement activists or organizations. Indeed, technological innovations are unlikely to spread just because they are "nifty." The limited popularity of the Segway Personal Transporter illustrates that, in the absence of sustained collective actions to promote them, new technologies are unlikely to catch the imagination of the public or diffuse widely (Rao 2009).

Second, the mobilization effort has to involve contentious politics and has to follow at least one of these pathways: an attempt to influence the policymaking process, an attempt to create consumer demand, or an attempt to change industrial sectors' rationales. If the collective actions do not follow at least one of these pathways, they are unlikely to be successful. Moreover, the more pathways that are pursued by activists and movement organizers, the more likely it is that the movement's influence will be significant. For example, noncontentious collective actions that stop at raising awareness of a certain issue are not likely to have a major impact on market formation.[29] Consider the environmental movement's attack on the fishing industry through its dolphin-safe tuna campaign in the United States, considered by many researchers to be quite successful. Although this campaign started with efforts to raise awareness of the fishing industry's negative consequences on dolphin populations, it became influential when it began concentrating on lobbying policymakers, targeting consumers through protests and boycotts, and changing some large corporations' perceptions about their responsibility to protect wildlife (Soule 2009).

Third, social movement mobilization efforts have to exist in tandem with a somewhat favorable social context and advantageous natural resources. Indeed, social movements' ability to shape industrial activities depends not only on their level of mobilization and on their chosen pathways of influence, but also on external factors such as the availability of influential political allies, the accessibility of mass media, and the presence of a sympathetic public opinion. This ability also depends on the types and quantities of natural resources that are available for specific economic activities. For example, anti-biotechnology mobilizations have impacted the German biotech industry much more than its U.S. counterpart. This is because anti-biotech activists in Germany, unlike those in the United States, have numerous political allies, interact with a mass media that is favorable to framing biotechnology experiments as "Frankensteinian," and encounter a public opinion that is unsympathetic to eugenics-like research (Rao 2009).

One promising direction for future research on social movements' impact on industry creation and development is to examine how the environmental movement contributes to other industries' growth. Indeed, it is likely that the processes through which the environmental movement has shaped the development of the wind energy industry are very similar to processes in other renewable energy industries. An interesting note is that Germany and Spain are world leaders not only in wind power but also in solar photovoltaics (PVs), and that the market for PVs in these countries is being driven by feed-in policies similar to those used for wind power.[30] Another interesting note is that German and Danish manufacturers dominate not only the global market for wind power technology but also the market for biomass technology.[31]

It is also likely that the environmental movement has a significant impact on other industrial sectors that are major contributors to air pollution and global climate change—for example, the automobile industry. In California, the same environmentalist ethic that motivated many pioneers to start up wind power companies during the 1970s and 1980s is also motivating many electric-car supporters and entrepreneurs nowadays. Not surprisingly, grass-roots organizations to support the development of electric-car technology emerged first and foremost in California.[32] Finally, it is worth exploring how other social movements influence either emerging or established industries; for example, the labor and social justice movements' impact on the mining and apparel industries.

Implications for Research on the Energy Sector

The book also has a few implications for future analyses of the electricity sector. An important question for future studies is whether or not the environmental movement will continue to play an important role. One possibility is that the environmental movement's role will become less and less significant, given that wind energy technology has matured and the price of wind power has decreased considerably. Indeed, some of those interviewed for this research argued that while during "the old days" most people in the industry were environmental activists—or, as one author put it, "ponytails"—the industry is now dominated by energy professionals who have little or no background as environmentalists.[33] Given that the wind energy industry is expected to directly employ almost 330,000 people in the European Union alone by 2020, it is likely that many new employees will be attracted to the industry because of its prospects for growth and not because of a calling to do what is right for the environment.[34] Thus, governmental support for wind energy could increase in some countries because the industry will create jobs, rather than because of its environmental benefits.

Additionally, global fears of "peak oil"—the point in time when the maximum rate of global petroleum extraction is reached—could accelerate the growth of the wind energy industry. Exact estimates of peak oil are difficult to make; some oil industry experts argue that peak oil production will be reached in 2010, while others argue that it will be reached in 2020 or later.[35] However, according to a U.S. Department of Energy report titled "Peaking of World Oil Production: Impacts, Mitigation, and Risk Management" (known as the Hirsch Report), "viable mitigation options exist on both the supply and demand sides, but to have a substantial impact, they must be initiated more than a decade in advance of peaking."[36] This means that, even if the optimistic scenarios of peak oil are valid, governments must act soon to avoid major disruptions in oil-thirsty sectors such as the transportation sector.

A number of scholars and journalists reinforce the argument that it is time to get serious about energy alternatives because a peak in global oil production is imminent.[37] Even former oil-men have been speaking out on the issue of peak oil: the best-known is T. Boone Pickens, who has called for the construction of more nuclear power plants, the use of natural gas to power the United States' transportation systems, and the promotion of alternative energy. In 2007, Pickens announced a plan to build what at the time was the largest proposed wind farm in the world (4 GW of capacity), and the next year he called on the government to support building power transmission lines to connect large wind turbine farms in the Great Plains to the power grid.[38] By mid-2009, however, due to ongoing problems in the capital markets, Pickens announced that he would cancel plans to build the 4 GW wind farm but would instead build a number of smaller wind farms.[39]

Concern about decreasing oil supplies has already motivated some oil and gas companies to invest in wind farms, and other companies have plans to diversify their energy portfolios by investing in wind power. As noted in chapter 5, in 2008 BP Alternative Energy entered into a fifty-fifty joint venture with Clipper Windpower to build the Titan wind project, a 5.05 GW wind energy project based in South Dakota.[40] Similarly, the Norwegian oil and gas company StatoilHydro announced in 2008 that it would build an offshore wind farm in the United Kingdom, and it started collaborating with other companies to develop deep-sea wind turbines known as the Hywind and Sway turbine projects.[41]

Another possibility, however, is that the environmental movement's role in the future will be as important as, if not even more important than, in the past. This is because the electricity sector will have to experience a colossal transformation to address the ever-more-pressing global climate change problem. Although it is very likely that the wind energy industry will continue to grow considerably even without assistance from the environmental movement, the movement's support could make the difference between a moderate and a fast rate of growth. An increase from slightly over 1 percent of global electricity consumption derived from wind power in 2008 to 12 percent in 2020, as the World Wind Energy Association has predicted, would be—to paraphrase Neil Armstrong—"a giant step for mankind." But an increase of this magnitude will not happen without major battles between environmental groups and renewable energy industries on one side, and fossil-fuel and nuclear energy industries on the other side. If environmentalists cannot convince governments to adopt and implement strong pro–renewable energy policies, create consumer demand for clean energy, and persuade electric utilities that building wind farms is a business opportunity—not a quixotic enterprise—this rapid growth will not be possible.

There are already signs that the conflict is heating up on multiple fronts. The nuclear energy industry is hoping for a renaissance in some countries. Nuclear energy supporters argue that nuclear power is carbon-free—or at least less polluting than coal—and, unlike oil, does not come from countries that harbor terrorists. Consider the following argument used in a *Forbes* article: "The average power plant is a little over 1,000 megawatts and the uranium fuel to run it for a year could fit in a single rail car. Compare that with coal, which would require an entire train filled with coal every day.

Uranium also comes from friendly places like Canada and Australia with little geopolitical risk."[42] In the United States, the licenses for many reactors have been extended from forty to sixty years, and seventeen license applications to build twenty-six new nuclear reactors have been advanced since mid-2007.[43] In a move that angered many environmentalists, in February 2010 President Obama announced $8 billion in loan guarantees to build the first new plant on U.S. soil in nearly thirty years. Moreover, according to Senate Minority Leader Mitch McConnell, "Senate Republicans support building 100 new [nuclear] plants as quickly as possible."[44]

In the United Kingdom, the government gave the green light in 2008 for a new generation of nuclear power stations, and some utilities have announced plans to build over 12 GW of new nuclear capacity.[45] While a few environmentalists are cautiously supporting new nuclear power to combat global climate change, most groups and activists intend to gear up for battle if new nuclear power plants indeed begin construction. For example, the executive director of Greenpeace UK commented on the government's decision: "This is bad news for Britain's energy security and bad news for our efforts to beat climate change."[46] Greenpeace successfully challenged an earlier U.K. government review backing nuclear power and is likely to challenge it again in the High Court and to organize vigorous protests. And, as in the past, environmental groups will not only say "no to nuclear power" but also "yes to wind" and other renewables.

The coal industry is also trying to adapt to a new political and social reality that makes coal increasingly unpopular. In the United States, the coal industry claims that coal is the cheapest energy source and that future advances in "clean coal" technology will greatly reduce coal's negative environmental impacts. The industry also argues that coal is a domestic resource, and it lobbies hard against any attempts to regulate greenhouse gas emissions. According to the American Coalition for Clean Coal Electricity (ACCCE)—a partnership of the industries involved in producing electricity from coal—"coal plays an important role in meeting our energy demands today. But some people ask if we can use coal to meet future energy needs and still achieve the commitment of reducing greenhouse gas emissions in response to climate change concerns? In a word—Yes! And, just as it has been in the past, technology and American ingenuity will provide the answers on how to get this job done."[47]

Environmentalists are not convinced. One author's statement sums up many environmental organizations' position: "Clean coal: Never was there an oxymoron more insidious, or more dangerous to our public health. Invoked as often by the Democratic presidential candidates as by the Republicans and by liberals and conservatives alike, this slogan has blindsided any meaningful progress toward a sustainable energy policy."[48] The environmental movement has already mobilized against coal power plants and has proposed a halt in construction of new coal-fired power plants. In addition to a number of dynamic organizations such as the Union of

Concerned Scientists, Rainforest Action Network, and the Energy Action Coalition, prominent scientists and politicians such as James Hansen and Albert Gore have also called for a coal moratorium. As James Hansen, director of NASA's Goddard Institute for Space Studies, stated: "There are long lists of things that people can do to help mitigate climate change. But for reasons quantified in my most recent publication, a moratorium on coal-fired power plants without carbon capture and sequestration is by far the most important action that needs to be pursued. It should be the rallying issue for young people."[49]

There are already some signs that the coal moratorium campaign is gaining momentum. The first de facto governmental moratorium on new coal plants in the United States (known as the "Schwarzenegger clause") took effect for investor-owned utilities in California in 2007.[50] A nationally representative opinion poll conducted in 2007 found that 75 percent of Americans would "support a five-year moratorium on new coal-fired power plants in the United States if there were stepped-up investment in clean, safe renewable energy—such as wind and solar—and improved home energy-efficiency standards."[51] It is estimated that, of the fifty-nine proposed coal power plants that were cancelled, abandoned, or put on hold in 2007, concerns about global warming played a major role in fifteen cases.[52] In 2008, the European Parliament's Environment Committee voted to support a limit on carbon dioxide emissions for new coal plants built in the European Union after 2015, a limit that is similar to the "Schwarzenegger clause."[53] Finally, in 2008 Henry Waxman, chair of the U.S. House Committee on Government Oversight and Reform, introduced H.R. 5575: Moratorium on Uncontrolled Power Plants Act of 2008.[54] Not surprisingly, the coal and oil lobbyists have fought this proposal tooth-and-nail—and, according to environmental groups such as Greenpeace, Friends of the Earth, and Public Citizen, so far they have succeeded.[55]

So, to the question, "Will the world be able to satisfy its growing demand for electricity and address the global climate change problem?" is the answer blowing in the wind? Will wind energy be an important part of the solution to the current crisis? Maybe; but it is likely that environmentalist winds of change, not only the atmospheric winds, will continue to shape the wind power industry's evolution. Environmentalists like to say that just as the Stone Age did not end because we ran out of stone, so the Fossil Fuel Age will not end because we run out of fossil fuels. We can certainly hope that's true, but we can also do our part to make sure that they are right.

Notes

Introduction

1. "Chorus of Congressional Voices Approves $100 Million Windbill," *Washington Post*, August 27, 1980, Metro section, C6.

2. "Huge Windmills Will Spin 'Juice' into 1,500 Homes," *Christian Science Monitor*, December 10, 1980, Financial section, 11.

3. Per Krogsgaard, head of BTM-Consult, quoted in a report produced by the German Wind Energy Association [*Bundesverband WindEnergie*], "Wind Energy Market 2007/2008," 10.

4. According to the Global Wind Energy Council, "this would represent an addition of 146 GW in 5 years, equaling an investment of over 180bn EUR (277 bn US$, both in 2007 value)" ("Global Wind 2007 Report," http://www.gwec.net/index.php?id=90 [accessed May 2010], 11).

5. Lester Brown noted in 2003 that "advances in wind turbine design since 1991 allow turbines to operate at lower wind speeds, to harness more of the wind's energy, and to harvest it at greater heights—dramatically expanding the harnessable wind resource. Add to this the recent bullish assessments of offshore wind potential, and the enormity of the wind resource becomes apparent. Wind power can meet not only all U.S. electricity needs, but all U.S. energy needs" ("Wind Power Set to Become World's Leading Energy Source," Earth Policy Institute, Plan B Updates, http://www.earth-policy.org/Updates/Update24.htm).

6. See www.londonpress.info/lps/article.asp?uniqueid=5857&category=environment; or http://edition.cnn.com/BUSINESS/programs/yourbusiness/stories2001/wind.turbine.

7. *Science* 308, no. 5727 (June 3, 2005): 1406.

8. See www.londonpress.info/lps/article.asp?uniqueid=5857&category=environment; or http://edition.cnn.com/BUSINESS/programs/yourbusiness/stories2001/wind.turbine.

9. See, for example, http://dorgan.senate.gov/about/northdakota/ (accessed July 2007). See also http://www.awea.org/projects/ (accessed July 2007).

10. See the report "IEA Summary: Deploying Renewables," http://www.iea.org/Textbase/npsum/DeployRenew2008SUM.pdf (accessed December 2009).

11. Werner Zittel, quoted in an Energy Watch Group report, http://www.folkecenter.net/mediafiles/folkecenter/awards/03_081227_Wind_Power_Report_Press_release_London.pdf (accessed March 2009); emphasis added.

12. See MacKay (2008, 189).

13. Matthew Wald, "It's Free, Plentiful and Fickle," New York Times, December 28, 2006.

14. Energy Watch Group, "Wind Power in Context—A Clean Revolution in the Energy Sector," a December 2008 report, http://www.energywatch-group.org/fileadmin/global/pdf/2009-01_Wind_Power_Report.pdf (accessed March 2009); emphasis in original.

15. See Environmental Literacy Council, http://www.enviroliteracy.org/subcategory.php/21.html (accessed May 2007); and World Coal Institute, http://www.worldcoal.org/pages/content/index.asp?PageID=21 (accessed May 2007).

16. See, for example, http://www.britannica.com/eb/article-10429/history-of-technology.

17. See Alex Gabbard, "Coal Combustion: Nuclear Resource or Danger," http://www.ornl.gov/info/ornlreview/rev26-34/text/colmain.html (accessed May 2010). It has also been estimated that in the year 2000 alone about twelve thousand metric tons of thorium and five thousand metric tons of uranium were released worldwide from burning coal (Gordon J. Aubrecht, "Nuclear Proliferation through Coal Burning," http://www.physics.ohio-state.edu/~aubrecht/coalvsnucMarcon.pdf#page=8 [accessed May 2010]).

18. For example, it is estimated that more than 90 percent of greenhouse gas emissions in the United States comes from the combustion of fossil fuels (U.S. Environmental Protection Agency, Inventory of U.S. Greenhouse Gas Emissions and Sinks: 1990–1998 Rep. EPA 236-R-00-01 [Washington, D.C.: U.S. Environmental Protection Agency, 2000], http://www.epa.gov/globalwarming).

19. The actual formula for calculating the amount of power is $P = 0.5 \times \rho \times A \times V^3$, where P = power in watts; ρ = air density in kilograms per cubic meter; A = rotor swept area in square meters; and V = wind speed in meters per second (American Wind Energy Association, http://www.awea.org/faq/windpower.html [accessed September 2007]).

20. For comparison, most cars run less than five hundred hours per year (Righter 1996, 183).

21. For example, Vestas—the number one wind turbine manufacturer in the world—experimented with the Darrieus turbine at the end of the 1970s, and the U.S. Department of Energy invested $28 million for the development of Darrieus technology between 1974 and 1985 (Gipe 1995, 85). However, the Danish concept was adopted in the early 1980s by most of the major wind turbine companies: Vestas, Nordtank, Bonus, and Micon.

22. For a detailed overview of wind turbines, see Hansen (2005).

23. See Hansen (2005).

24. American Wind Energy Association, "The Economics of Wind Energy," http://www.awea.org/pubs/factsheets/EconomicsOfWind-Feb2005.pdf.

25. American Wind Energy Association, http://www.awea.org/faq/cost.html.

26. "Capacity factor" is defined as the actual amount of power produced over time, divided by the power that would have been produced if the turbine operated at maximum output 100 percent of the time (American Wind Energy Association, http://www.awea.org/faq/wwt_basics.html [accessed June 2008]).

27. Some turbines use pitch control—they turn the blades in order to control the amount of power used. Others use stall control, a special aerodynamic design of the rotor blade to regulate the power of the wind turbine. See, for example, Söder and Ackermann (2005, 35).

28. For example, one study found that the hourly correlation between the Texas coast and the rest of the state is essentially zero (Kirby 2007).

29. American Wind Energy Association, Fact Sheets, http://www.awea.org/resources/resource_library/#FactSheets (accessed April 2010).

30. See Sovacool (2008).

31. See Erickson, Johnson, and Young (2005).

32. According to the American Wind Energy Association (AWEA), global sales of the international wind turbine industry reached around US$4 billion in 2000 (www.awea.org/faq/global2000.html). The AWEA estimates that even under a modest growth scenario, annual sales could reach US$26 billion by 2010 (www.awea.org/AWEA_SWT_Market_Study_6-05.pdf).

33. The semistructured interviews were conducted between September 2005 and May 2009. The interviews lasted between fifteen minutes and two hours and were conducted over the telephone or in person. The interviewees were located in the following countries: Canada, Denmark, Germany, Spain, the United Kingdom, and the United States. Some interviews were used only for background information; many were quoted to illustrate opinions, strategies, or facts.

Chapter 1

1. The article, however, did not mention that high growth rates of wind power are based on a small base, and small industries can achieve high growth more easily than larger, long-established industries.

2. "Wind Power Fastest Growing Energy Source Ready to Displace Coal, Slow Climate Change," Worldwatch Institute, August 14, 1996, http://www.worldwatch.org/node/1598 (accessed July 2008).

3. For example, scholars writing about the growth of the flat-panel display industry in Japan argue that "there was nothing intrinsic about Japan, Japanese management style, or any other Japanese business, academic or government institution that uniquely destined Japan to serve as the global center of the industry" (Murtha, Lenway, and Hart 2001, 38).

4. The United States spent about twenty times more, while Germany spent about five times more than Denmark (Heymann 1998). Other studies come to similar conclusions when comparing the development of the wind energy industry in countries such as the United States and Denmark (Est 1999); Germany, the Netherlands, and Sweden (Johnson and Jacobsson 2000); and the Netherlands and Denmark (Boon 2008).

5. Personal interview with a representative of the Texas Renewable Energy Industries Association, November 2007.

6. Sorenson and Audia (2000) develop the metaphor of the pollination process without describing the nature of "the wind." However, they mention in passing that the role of the wind can be played by policymakers who start the pollination process in a new region by "recruiting one or more success-ful companies to the region that can 'fertilize' the area" (457).

7. An analysis of the structure of industrial networks is relevant for the early stages of entrepreneurial activity, when industry standards are not yet formed, but less informative for the later stages of industry growth, when the technology matures and standard technical solutions are widely adopted.

8. The Public Utility Regulatory Policies Act (PURPA) is often cited as a classic example of a policy that broke utilities' local monopolies on electricity generation and stimulated the growth of independent power producers—and, implicitly, the wind energy industry—by encouraging long-term power purchase contracts with utilities. PURPA policy mandated that utilities pur-chase power generated by small electricity generators and that they pay the independent power producers at the "avoided cost"—the cost that the utility would have to pay to generate the electricity itself (Redlinger, Andersen, and Morthorst 2002). However, since the implementation of the policy was left to individual states and the definition of avoided cost varied greatly by state, PURPA's nationwide success has been limited. In California, where the Public Utilities Commission ordered the institution of standard contracts between utilities and qualifying independent electricity generators to imple-ment PURPA, the absence of a coherent federal tax policy on wind energy, and abuse of the state wind energy income tax, contributed to a "wind rush" in the early 1980s that "would help get the machines in the ground, but could not ensure that, once installed, they would continue to operate" (Asmus 2001, 75).

9. Social movement scholars have emphasized the crucial difference between social movements' purposive and unintended consequences. For more details, see Giugni, McAdam, and Tilly (1999).

10. For example, while the peace movement has a low chance to impact policy because it targets an issue area that is very difficult to change, the environmental movement has a relatively high chance to impact policy because it addresses "valence issues" (Giugni 2007).

11. It is estimated that about fifty thousand small wind-electric systems operated in 1950 in the rural areas of the United States (Righter 1996, 99).

12. *New York Times*, February 13, 1971, 27. Quoted in Righter (1996, 153).

13. For example, in opening the hearings for this bill, Senator Frank Moss of Utah recognized that "fossil fuel electric power plants rank with the automobile as the Nation's worst polluter" (Righter 1996, 154).

14. See Righter (1996, 151).

15. In 1980 the Union of Concerned Scientists published a report that stated that wind is "perhaps the most ecologically benign source of electric power" and that "in contrast to conventional methods of power generation, wind energy conversion is highly favorable from an environmental standpoint, posing no major societal risks" (Kendall and Nadis 1980, 156–57).

16. The Western Montana G & T survey found that 23 percent supported wind turbines; 21 percent, solar plants; 19 percent, hydroelectric dams; 13 percent, natural-gas combustion turbines; 10 percent, wood and municipal-waste burning plants; 5 percent, nuclear plants; and 3 percent, coal plants (Smeloff and Asmus 1997, 204).

17. Wind turbine manufacturers have frequently argued that wind turbines contribute to national energy independence and national security. Some manufacturers even name their turbines to evoke national energy independence; for instance, Clipper Windpower, a start-up turbine manufacturer, is producing a 2.5 MW turbine named Liberty.

18. See Righter (1996, 203). Closer to the present, President George W. Bush stated that "the best way to break the addiction to foreign oil is through new technology" and decided to increase funding for wind energy research and to expand access to Federal lands for wind energy development "in order to dramatically increase the use of wind energy in the United States" (State of the Union: The Advanced Energy Initiative, January 31, 2006, http://www.whitehouse.gov/news/releases/2006/01/20060131-6.html [accessed June 2006]).

19. Three mechanisms lead to isomorphic change in organizations: "1) coercive isomorphism that stems from political influence and the problem of legitimacy; 2) mimetic isomorphism resulting from standard responses to uncertainty; and 3) normative isomorphism, associated with professionalization" (DiMaggio and Powell 1983, 150). Research has also emphasized that organizations are located in fields, or "communities of organizations that partake of a common meaning system and whose participants interact more frequently and fatefully with one another than with actors outside the field" (Scott 2002, 129). Organizational fields are characterized by homogenizing institutional pressures (Meyer and Rowan 1977; DiMaggio and Powell 1983; DiMaggio 1991), as well as by conflicting institutional processes (Hoffman and Ventresca 2002; Levy and Rothenberg 2002). Recent research has also emphasized that organizational fields are often overlapping and nested—see, for example, Vasi (2007).

20. Survey research on determinants of pro-environment behaviors such as green consumption or green-energy conservation suggests that the environmental movement could create demand for green energy by changing individuals' personal values and by increasing the salience of their "environment identity" (Granzin and Olsen 1991; Mainieri et al. 1997; Stets and Biga 2003; Poortinga, Steg, and Vlek 2004; Carlisle and Smith 2005).

21. See Sine and Lee (2009).

22. See Thornton and Ocasio (1999); Haveman and Rao (1997); and Rao, Monin, and Durand (2003).

23. For a description of the hybrid figure of the activist-entrepreneur in technology-oriented movements, see Hess (2007).

24. According to McAdam, Tarrow, and Tilly (2001, 24), mechanisms are "a delimited class of events that alter relations among specified sets of elements in identical or closely similar ways over a variety of situations."

25. McAdam, Tarrow, and Tilly (2001, 25–26).

26. For example, the most extensive study that examines coverage in the mass media of environmental protest activities is limited to eight countries (Rootes 2003).

27. See Smith and Wiest (2005); and Dalton (2005).

28. See Redlinger, Andersen, and Morthorst (2002, 171).

29. Given this study's focus on the environmental movement's impact, data collection problems prevented the analysis of all countries.

30. An alternative measure of the dependent variable includes the total number of megawatts from wind power, transformed with the natural logarithm to stabilize skew in the data. For more information on alternative measures and models, see Vasi (2009).

31. The year 2003 was chosen for consistency with other measures; measures of GDP per capita from different years over the last decade were also used and had similar effects.

32. The year 2003 was selected for consistency with other variables; measures of the electricity sector's pollution from other years were also used and did not produce significantly different results.

33. Other measures of the strength of domestic environmental groups are unfeasible for a large cross-national study. One such measure is self-reported per capita membership in nongovernmental environmental organizations; yet the most extensive study that examines membership in environmental organizations considers only fifty countries (Dalton 2005). Another measure is a count of environmental protest activities reported in the mass media; yet the most extensive research on environmental protests to date examines only eight countries (Rootes 2003).

34. To check for robustness of findings, a second measure of the presence of political allies used the weighted "value of environmentalism" scale, which ranks countries in terms of their participation in twelve international environmental-protection conventions in the period 1970–1990 (Dietz and Kalof 1992). This alternative measure produced similar results. These measures assume that participation in international environmental agreements and a high number of intergovernmental environmental organizations can be used as proxy measures for political allies. While these measures are not ideal, they are the best available proxy measures for political allies of the environmental movement.

35. The variable adoption of FITs is not included in the correlation analysis because it was coded as a dummy—a variable that has only two values.

36. The problem of multicolinearity is that two independent variables can be correlated to such a high degree that their effects on the dependent variable are indistinguishable.

37. In contrast to FITs, other policies such as investment incentives can inflate the costs for wind projects and can result in opportunistic behavior from investors who seek tax shelters rather than actual electricity production. Production incentives eliminate the temptation to inflate project costs,

but their long-term unpredictability due to various causes—from political whim to government budget cutbacks—frequently has a devastating impact on wind energy projects (Redlinger, Andersen, and Morthorst 2002).

Chapter 2

1. See the German Wind Energy Association website http://www.wind-energie.de/en/wind-energy-in-germany/ (accessed July 2008).

2. See the Earth Policy Institute website http://www.earth-policy.org/Indicators/Wind/2008_data.htm (accessed July 2008).

3. The total installed nominal capacity from wind power in Denmark was 3.16 GW, and in Spain it was 16.74 GW ("World Wind Energy Report 2008," http://www.wwindea.org/home/images/stories/worldwindenergyreport 2008_s.pdf [accessed March 2008]).

4. For more information on the number of jobs created by the wind energy industry in Europe, see Blanco and Rodrigues (2009).

5. See the European Wind Energy Association report "Wind Power Installed in Europe by End of 2008 (cumulative)," http://www.ewec2009.info/fileadmin/ewec2009_files/documents/Media_room/European_Wind_Map_2008.pdf (accessed July 2009).

6. For a comprehensive review of FITs, see Mendonça (2007).

7. REN21 [Renewable Energy Policy Network for the 21st Century], "Renewables 2007: Global Status Report," www.ren21.net/pdf/RE2007_Global_Status_Report.pdf, 23–24.

8. See HM Treasury, "Stern Review on the Economics of Climate Change," 2006, 367.

9. Activist from the environmental group Rheintal-Aktion, quoted in Joppke (1993, 97).

10. Some social movement scholars argue that many of the antinuclear groups created in Germany during the 1970s had an identity that was distinct "from the environmental movement at large" (Wagner 1994, 272). Although the anti–nuclear energy and pro-environment mobilizations are sometimes analyzed as separate social movements, I consider antinuclear protests as campaigns of the environmental movement because environmental groups are often the main organizers and because many of these protests are "miscible mobilizations," or "mobilization efforts by movements with compatible ideologies and shared activist communities and SMOs [social movement organizations]" (Vasi 2006, 137). This, of course, does not prevent groups that normally have a low level of miscibility, such as conservative and progressive groups, from forming temporary alliances against nuclear power—as was the case in the protests at Wyhl (Joppke 1993, 100).

11. See Rucht and Roose (2003).

12. See "Heated Debate in Bonn over Nuclear Waste Transport Sabotage," *Deutsche Presse-Agentur*, February 27, 1997.

13. The "anti-Castor" protests were named after the type of containers used for transporting nuclear waste. These protests had broad support from a wide range of Germans: farmers, retirees, clergy. In one show of solidarity,

for example, a local fire department refused to provide water for water cannons that police planned to use against demonstrators ("Germany's Greens See Red over Nuclear-Waste Storage," *Christian Science Monitor*, March 6, 1997).

14. The constant decline in public support for nuclear energy due to the antinuclear mobilizations accelerated after the nuclear accident at Chernobyl—see Kolb (2007, 211).

15. As the federal minister for research and technology, Hans Matthöfer stated, "We need economic growth if we want to have full employment, and this is not possible without new nuclear capacity"—see Joppke (1993, 100).

16. For example, Roland Vogt and Petra Kelly left the SPD and became key figures in building the Green Party (Joppke 1993, 238).

17. This is an ambitious goal since in 2004 Germany was the fourth-largest generator of nuclear power in the world and had seventeen operating nuclear power plants (http://www.eia.doe.gov/emeu/cabs/Germany/Electricity.html).

18. See Storchmann (2005).

19. See Jacobsson and Lauber (2005, 130).

20. Like all national coal industries, the German coal industry is heavily subsidized. The annual subsidy increase from 1959 to 1970 was about 5 percent, and the growth in the 1970s was about 15 percent per year. However, partly because of increasing opposition from the Green Party and environmental groups, annual growth in real subsidies slowed considerably in the 1980s to 4.8 percent per year, and from the late 1980s real subsidies declined by an average of 5.7 percent per year (Storchmann 2005, 1490).

21. This information is based on an analysis of two major news sources, *Bild der Wissenschaft* and *Der Spiegel*—see Beuermann and Jäger (1996, 192).

22. See Rucht and Roose (2003).

23. See Rucht and Roose (2003, 90, 97).

24. See the Environmental News Service website http://www.ens-newswire.com/ens/may2004/2004-05-28-02.asp (accessed August 2008).

25. For example, a 2002 poll showed that 86 percent of German citizens were in favor of increasing wind's contribution to the energy mix. Another survey from 2003 showed that 76 percent of Germans considered nuclear and coal-fired power plants to "spoil the landscape," and only 27 percent thought the same thing about wind turbines (European Wind Energy Association 2003).

26. See Gipe (2004).

27. Personal interview with a Greenpeace Germany organizer, November 2008.

28. See Alenka Burja, "Energy Is a Driving Force for Our Civilisation: Solar Advocate, An Interview with Hermann Scheer," Nordic Folkecenter for Renewable Energy, http://www.folkecenter.net/default.asp?id=8481 (accessed May 2010).

29. Northern Germany has the best wind resources in Germany and resembles Denmark, which had the most advanced wind energy industry in Europe in the 1980s.

30. Under the 250 MW program, Germany paid 6 pfennigs (US$0.035) per kilowatt-hour of generation from selected wind turbines (Jacobsson and Lauber 2005, 133).

31. See Jacobsson and Lauber (2005); and Lauber and Mez (2006).

32. Personal interview with a member of the World Wind Energy Association, October 2008.

33. See Lauber and Mez (2006, 106).

34. Because of the local ownership of many wind projects, NIMBY opposition was minimized and further development was encouraged. According to one newspaper article, Germany has very few anti–wind power groups, and they are very weak. For example, despite encouraging people to refuse to sell their land for wind farms and challenging planning applications, groups such as the Landscape Protection Association have made little impact on the rate of windmill construction ("Society: Alternative energy, 'Where There's a Mill...'" *Guardian*, April 29, 1998).

35. See Toke and Lauber (2007).

36. Another goal of this act was to stimulate local economic growth. The wind energy industry currently provides over sixty-four thousand people with jobs, such that wind energy has overtaken coal mining as a major employer (German Wind Energy Association, "A Clean Issue—Wind Energy in Germany," http://www.wind-energie.de/fileadmin/dokumente/English/Broschueren/BWEImageEngl_2006.pdf [accessed May 2010]).

37. Act on Granting Priority to Renewable Energy Sources [Renewable Energy Sources Act], section 1, page 5.

38. See German Wind Energy Association, "A Clean Issue," 10.

39. See Lauber and Mez (2006, 107).

40. Personal interview with a member of the World Wind Energy Association, October 2008.

41. See "Germany Rethinks Renewables Policy," *Generation Week*, September 25, 1998, 38.

42. Personal interview with a Greenpeace Germany organizer, November 2008.

43. A detailed list of the organizers includes Bundesverband Erneuerbare Energien, IG Metall, Eurosolar, Bund der Energieverbraucher, Bundesverband Mittelständische Wirtschaft, Bundesverband Deutscher Wasserkraftwerke, Greenpeace, Bundesverband Solarindustrie, Bundesverband WindEnergie, Bund für Umwelt und Naturschutz Deutschland, Wirtschaftsverband Windkraftwerke, Deutsche Gesellschaft für Sonnenenergie, Deutsche Umwelthilfe, Fachverband Biogas, Energie-Agenturen Deutschland e.V., Grüne Liga, Naturschutzbund, Robin Wood, Unternehmensvereinigung Solarwirtschaft, Bundesverband Bürgerinitiativen Umweltschutz, Geothermische Vereinigung, Solarförderverein, the European Renewable Energies Federation, and the World Wildlife Fund (Michaelowa 2005, 198).

44. Personal interview with a member of the World Wind Energy Association, October 2008.

45. See Jacobsson and Lauber (2005, 136).

46. Personal interview with a member of the World Wind Energy Association, October 2008.

47. For a similar argument, see Wagner (1994).

48. See Schreurs (2003).

49. For example, Germany has some of the strictest environmental regulations in the world and was the first to adopt and implement domestic environmental policies centered on the precautionary principle (Eckersley 2004).

50. See Alenka Burja, "Energy Is a Driving Force for Our Civilization: An Interview with Hermann Scheer." http://www.folkecenter.dk/en/articles/HScheer_aburja.htm (accessed September 2008).

51. In 2007, Denmark followed an incentive system introduced in June 2004 with a premium of kr 0.10/kWh paid on top of the market price for twenty years. The system includes a cap of kr 0.36/kWh for turbines installed before the end of 2004 plus an allowance of kr 0.023/kWh (International Energy Agency, "IEA Wind Energy 2007 Annual Report," chap. 13, "Denmark,"http://www.ieawind.org/AnnualReports_PDF/2007/CountryChapters/Denmark.pdf [accessed November 2008]).

52. For a comprehensive review of the Danish policies, see K. Nielsen (2005).

53. See Jamison et al. (1990, 90).

54. H. Nielsen (2006, 216).

55. See Jamison et al. (1990, 104).

56. See Erik Grove-Nielsen, "A Personal Story in Photos Told by Early Blade-Manufacturer Erik Grove-Nielsen," http://www.windsofchange.dk/WOC-75-77.php (accessed November 2008).

57. One of the first pro–renewable energy policies was adopted in 1979 when the Danish parliament passed legislation that secured a state grant of 30 percent of the cost of a new wind turbine. The subsidy was reduced to 20 percent in 1985 and was eliminated in 1989 (K. Nielsen 2005, 110).

58. See the Tvind Internationale Skolecenter website http://www.tvind.dk/TextPage.asp?MenuItemID=55&SubMenuItemID=160 (accessed November 2008).

59. Grove-Nielsen, "A Personal Story in Photos."

60. Tvind Internationale Skolecenter, http://www.tvind.dk/TextPage.asp?MenuItemID=55&SubMenuItemID=160.

61. Grove-Nielsen, "A Personal Story in Photos."

62. Henrik Stiesdal, technical director for the wind turbine producer Bonus Energy, who visited Tvind for the first time in 1976, as quoted in Engineer—see Tvind Internationale Skolecenter, http://www.tvind.dk/TextPage.asp?MenuItemID=55&SubMenuItemID=160.

63. Tvind Internationale Skolecenter, http://www.tvind.dk/TextPage.asp?MenuItemID=55&SubMenuItemID=160.

64. Tvind Internationale Skolecenter, http://www.tvind.dk/TextPage.asp?MenuItemID=55&SubMenuItemID=160.

65. See the Organization for Renewable Energy website http://www.ove.org/index.php?la=eng&id=3 (accessed November 2008).

66. Organization for Renewable Energy, http://www.ove.org/index.php?la=eng&id=3.

67. A 2001 survey showed that 68 percent of the Danish population answered yes to the question "Should Denmark continue to build wind turbines to increase wind power's share of electricity production?" In contrast, 18 percent found the current level satisfactory and only 7 percent answered

that there were already too many wind turbines (European Wind Energy Association 2003).

68. See Gipe (2004).

69. See the 92 Gruppen website http://www.92grp.dk/inenglish/denmark_on_track.htm (accessed November 2008).

70. For example, the Danish Wind Turbine Owners' Association (DV) published *Naturlig Energi*, a magazine that not only educated the public about wind turbines but also benefited wind turbine engineers and manufacturers by providing vital monthly statistics about the functioning of wind turbines around the country.

71. As the minister of environmental protection argued in parliament in 1972, "Economic values must receive less importance and be supplemented by other values, especially ecological"—see Jamison et al. (1990, 78).

72. K. Nielsen (2005, 109).

73. In its 2007 energy policy proposal A Visionary Danish Energy Policy 2025, the government adopted two important targets: increasing the share of renewable energy to at least 30 percent of energy consumption by 2025; and a doubling of publicly funded research and development and demonstration of energy technology to kr 1 billion annually from 2010 onward—see http://www.ieawind.org/AnnualReports_PDF/2007/CountryChapters/Denmark.pdf.

74. Personal interview with Henrik Stiesdal, February 2009.

75. Mario Ragwitz and Claus Huber, "Feed-in Systems in Germany and Spain—A Comparison," http://www.bmu.de/files/english/renewable_energy/downloads/application/pdf/langfassung_einspeisesysteme_en.pdf (accessed September 2008).

76. The ETA was directly involved in direct, violent actions against the Lemoniz nuclear reactor; in 1977 it carried out arson attacks and even organized a commando attack to blow up part of the installations—an attack that resulted in the killing of one ETA militant (Rüdig 1990, 139).

77. See Rüdig (1990, 339).

78. "Spain Says 'Adios' to Nuclear Power: Fourth European Country to Begin Phase Out," http://www.greenpeace.org/international/news/spain-adios-nuclear-31-06-06 (accessed August 2008).

79. Additionally, as Jiménez (2007, 360) notes, the "peripheral nationalisms favored decentralized organizational models and localism, making it difficult to develop stable coordinating structures on a state-wide basis."

80. Jiménez (2007, 375) estimates that the number of environmental groups in the late 1990s was around one thousand.

81. However, Jiménez (1999, 159) points out that many environmental groups have withdrawn from CAMA because they see it as a mere mechanism for legitimating the government's environmental policy.

82. The combined issues of clean energy and anti–nuclear power ranked as number one; when clean energy was considered on its own, it ranked as the third-most-important issue, after urban waste and water pollution (Jiménez 2007, 367).

83. According to Jiménez (2003, 190), Greenpeace participated in 103 protests, while both CODA and AEDENAT participated in 62 protests. Jiménez (2007, 375) also points out that AEDENAT was founded in 1976, CODA in 1979, and Greenpeace Spain in 1984.

84. See http://www.greenpeace.org/raw/content/espana/reports/energ-a-e-lica-terrestre-plan.pdf.

85. See http://www.greenpeace.org/international/news/spain-adios-nu-clear-31-06-06.

86. According to this Greenpeace report, the costs of conversion would be €120,000 million, to be spent over a twenty-five-year period (http://climate-change.suite101.com/article.cfm/greenpeace_report_on_spain.

87. See http://www.climateark.org/shared/reader/welcome.aspx?linkid=10 7751&keybold=renewablepercent20energypercent20carbonpercent20free.

88. Personal interview with a Greenpeace Spain organizer, September 2008.

89. Ibid.

90. Ibid.

91. Personal interview with Begoña Urien, general director of enterprise of the Department of Innovation, Enterprise, and Employment for the government of Navarra, October 2008.

92. For instance, a 2001 opinion poll showed that 85 percent of the Spanish public was in favor of wind power, up from 75 percent in 1998. In contrast, only 1 percent of the Spanish public was in favor of nuclear energy (European Wind Energy Association 2003).

93. Cynthia Graber, "Wind Power in Spain," *Technology Review*, S2, http://www.uprm.edu/aceer/pdfs/wind_power_spain.pdf (accessed May 2010).

94. The data on the year of adoption of FITs was obtained from Mendonça (2007). The data on the percentage of energy produced from wind power in 2005 was obtained from the European Commission, Directorate-General for Energy and Transport, "Renewable Energy Fact Sheet," http://www.energy.eu/renewables/factsheets/2008_res_sheet (accessed June 2009).

95. See the International Energy Agency website http://www.iea.org/country/m_country.asp?COUNTRY_CODE=FR (accessed June 2009).

96. The data on the percentages of energy produced from nuclear energy and hydropower in Sweden was obtained from the European Commission, Directorate-General for Energy and Transport, "Renewable Energy Fact Sheet," http://www.energy.eu/renewables/factsheets/2008_res_sheet (accessed June 2009).

97. See "Sweden Plans to be World's First Oil-Free Economy," *Guardian*, February 8, 2006, http://www.guardian.co.uk/environment/2006/feb/08/frontpagenews.oilandpetrol (accessed June 2009).

Chapter 3

1. See http://www.earth-policy.org/Indicators/Wind/2008_data.htm.

2. Although the United States had overtaken Germany in terms of total installed capacity at the end of 2008, the United States has much better wind resources and a much larger land surface than Germany.

3. Of course, the strong neoliberal ideological tradition present in these countries, particularly in the United States, partly accounts for the fact that they adopted RPS and not renewable energy FIT policies. For more on this, see Lauber (2005); or Paul Gipe, "Feed-in Tariffs in Britain: Ideological Break through," http://www.wind-works.org/FeedLaws/Great%20Britain/Feed-in TariffsinBritainIdeologicalBreakthrough.html (accessed November 2008).

4. Although the NFFO had a target of bringing 1.5 GW of new capacity from renewable energy by 2000, it had only achieved approximately 1 GW by 2002. Moreover, this policy was criticized for failing to establish significant U.K. renewable energy technology manufacturing (Connor 2005, 163–64).

5. See Gipe, "Feed-in Tariffs in Britain."

6. See "People-Power a Step Closer in Energy Bill," *Guardian*, October 31, 2008, http://www.guardian.co.uk/technology/2008/oct/31/renewable-energy-micro-generation-national-grid.

7. See Rootes (2003, 42); Rucht and Roose (2003, 95); and Jiménez (2003, 175).

8. See "Filthy Britain 'A Pollution Failure,'" *Observer*, February 3, 2002, http://www.guardian.co.uk/uk/2002/feb/03/anthonybrowne.theobserver (accessed May 2010).

9. Many U.K. politicians support strong climate change action and renewable energy but also nuclear power. Consider the following statement by Tony Blair: "We can meet our carbon dioxide emissions targets, but only if we are willing to think ahead and take tough decisions over new wind farms—and give serious consideration to nuclear power" ("How to Stop the Lights Going Out in a Dangerous World," *Times Online*, May 23, 2007, http://www.timesonline.co.uk/tol/comment/columnists/guest_contributors/article1826518.ece [accessed May 2010]).

10. For example, natural-gas power plants went from making a negligible contribution to electricity generation in the early 1990s to making the largest contribution to the country's electricity supply by 2000—almost 40 percent (Rowlands 2005, 70).

11. Rising Tide does not have a formal membership structure and emphasizes the importance of direct action. According to its mission statement, "We believe that public protest has always played a crucial role in movements for social change. We publicize campaigns and actions by groups in the network and encourage people to support them. We also provide access to networks and experienced activists" (http://risingtide.org.uk/about [accessed May 2010]).

12. The coalition asks the U.K. government to

1. Show leadership to help create a fair international deal by 2010 to keep global warming below the 2 degrees C danger threshold to protect people and the planet.
2. Take immediate practical action to deliver substantial, sustained annual emissions reductions ensuring that the UK meets its fair share of the international effort to keep global warming below 2 degrees C.

3. Provide poor countries with the resources they need to help them adapt to climate change and follow a low carbon development path (Stop Climate Chaos Coalition, http://www.stopclimatechaos.org/we-are [accessed May 2010]).

13. Mass demonstrations were also organized by the Stop Climate Chaos Coalition in December 2007 and 2008. Each time the organizers estimated that approximately ten thousand people participated ("Global Rallies Focus on Climate," *BBC News*, http://news.bbc.co.uk/2/hi/uk_news/7134060.stm; and "Climate Change Campaigners Rally," *BBC News*, http://news.bbc.co.uk/2/hi/uk_news/england/london/7768867.stm [accessed April 2009]).

14. See the Friends of the Earth website http://www.foe.co.uk/campaigns/climate/news/big_ask.html (accessed April 2009).

15. See the Friends of the Earth website http://www.foe.co.uk/campaigns/climate/success_stories/gov_climate_bill.html (accessed March 2009).

16. House of Commons Hansard Debates, October 28, 2008, http://www.publications.parliament.uk/pa/cm200708/cmhansrd/cm081028/debtext/81028–0021.htm (accessed April 2009).

17. See "Blair Shuns Yearly Targets to Reduce Carbon Emissions," *Independent*, November 15, 2006, http://www.independent.co.uk/news/uk/politics/blair-shuns-yearly-targets-to-reduce-carbon-emissions-424365.html (accessed March 2009).

18. See the Greenpeace website http://www.greenpeace.org/international/press/releases/court-major-blow-to-uk-coal-10092008 (accessed December 2008).

19. "Clean Energy 'Revolution' Not Delivering, Says Survey," *Sunday Herald*, February 28, 1999.

20. See "Nuclear Power Plants Get Go Ahead", *BBC News*, July 11, 2006, http://news.bbc.co.uk/2/hi/uk_politics/5166426.stm (accessed May 2010).

21. See "Nuclear Review 'Was Misleading,'" *BBC News*, February 15, 2007, http://news.bbc.co.uk/2/hi/uk_news/politics/6364281.stm (accessed May 2010).

22. See "Government Nuclear Talks Pointless, Say Green Groups," *Guardian*, September 7, 2007, http://www.guardian.co.uk/environment/2007/sep/07/nuclearpower.nuclearindustry (accessed May 2010).

23. Some of the environmental activists and writers who expressed certain support for nuclear power in the United Kingdom were James Lovelock, George Monbiot, and Mark Lynas. Globally, a nonprofit organization Environmentalists for Nuclear Energy states that currently "it gathers over 9,000 members and supporters, with 255 local correspondents in 60 countries, on five continents." According to its website, "Fossil fuels (oil, gas, coal) the dominant energy today, are being rapidly exhausted, and are the cause of wide scale pollution of our environment, while nuclear and renewable energies are much cleaner: they have absolutely no global effect, produce only very small amounts of waste that are easy to manage, don't affect the planet's climate, and these energy sources (renewable energies and clean nuclear energy), if well managed, are sustainable in the very long term" (EFN [Environmentalists for Nuclear Energy], http://www.ecolo.org/base/baseus.htm [accessed May 2010]).

24. See the Friends of the Earth website http://www.foe.co.uk/resource/press_releases/campaigners_call_for_renew_22042008.html (accessed December 2008).

25. See the Renewable Energy Association website http://www.r-e-a.net/info/rea-news/feed-in-tariff-campaign-could-push-renewables-the-uk-into-eu-premier-league (accessed November 2008).

26. Friends of the Earth, http://www.foe.co.uk/resource/press_releases/campaigners_call_for_renew_22042008.html.

27. Personal interview with a member of the Renewable Energy Association, November 2008.

28. Ibid.

29. Ibid.

30. Friends of the Earth, http://www.foe.co.uk/resource/press_releases/campaigners_call_for_renew_22042008.html.

31. See Toke and Lauber (2007).

32. See Paul Gipe, "British Feed-in Tariff Policy Becomes Law—Was Once Unthinkable," http://wind-works.org/FeedLaws/Great%20Britain/BritishFeed-inTariffPolicyBecomesLaw.html (accessed November 2008).

33. Personal interview with a member of the Renewable Energy Association, November 2008.

34. The fossil-fuel and nuclear energy industries' opposition was moderate because most of the electricity in the United Kingdom is produced from natural gas, which is much cleaner than coal, and from nuclear power, which portrays itself as "climate friendly." For example, in 2006 natural gas and nuclear power produced together approximately twice the amount of electricity produced from coal.

35. See Ryan Wiser and Galen Barbose, "Renewables Portfolio Standards in the United States: A Status Report with Data through 2007," Lawrence Berkley National Laboratory, http://eetd.lbl.gov/ea/ems/reports/lbnl-154e.pdf (accessed May 2010).

36. In 2007, the U.S. House of Representatives voted in favor of including an RPS as part of its energy bill. This bill would have established a 15 percent RPS by 2020, but the Senate energy bill did not include an RPS due to the uncertainty that the sixty votes needed to overcome a likely filibuster would have been secured (American Wind Energy Association, http://www.awea.org/legislative/#RPS [accessed December 2008]).

37. Production tax credits are designed to stimulate investment by providing tax credits to green-energy investors. They have been criticized mainly because, in order to take advantage of the tax credits, projects have to be financed with a great proportion of high-cost equity and a low proportion of low-cost debt (Redlinger, Andersen, and Morthorst 2002, 175).

38. See Tamplin and Cochran (1974).

39. However, as Joppke (1993, 135) has argued, the decline of the nuclear industry is attributable mainly to a combination of skyrocketing capital costs and declining demand for electricity.

40. See Union of Concerned Scientists, "Declaration on the Nuclear Arms Race," *Bulletin of Atomic Scientists*, March 8–10, 1978.

41. Membership in some organizations has fluctuated; for example, Greenpeace had approximately 250,000 members in 1980, over 2 million in 1990, and 250,000 in 2003. Similarly, the National Wildlife Federation had 818,000 members in 1980, approximately 1 million in 1990, and 650,000 in 2003 (Bosso 2005, 54).

42. See Lovins (1977).

43. See Kendall and Nadis (1980, 157).

44. The Crude Oil Windfall Profits Act of 1980 extended the credit to December 1985, when it was allowed to lapse for wind (Energy Information Administration, "Production Tax Credit for Renewable Electricity Generation," report, http://www.eia.doe.gov/oiaf/aeo/otheranalysis/aeo_2005 analysispapers/prcreg.html [accessed December 2008]).

45. Between 1978 and 1981, renewable energy R & D was US$ (1996)1,290 million, while fossil-fuel and nuclear energy R & D were US$ (1996)1,538 and US$ (1996)2,878 million respectively. For comparison, between 1982 and 1990, renewable energy R & D was US$ (1996)2,279 million, while fossil-fuel and nuclear energy R & D were US$ (1996)3,712 and US$ (1996)13,727 million respectively (American Physical Society, "R & D Priorities within the Department of Energy," report, http://www.aps.org/policy/reports/popa-reports/energy/doe.cfm [accessed December 2008]).

46. For a brief description of the New Alchemy Institute, see http://www .nature.my.cape.com/greencenter/newalchemy.html (accessed December 2008).

47. Paul Gipe, quoted in Righter (1996, 202).

48. The changes in political opportunity at the federal level due to the election of Ronald Reagan in 1982, and at the state level due to the election of Republican George Deukmejian as California's governor, led to expiration of the investment tax credit and slowed the development of the wind energy industry in California after the mid-1980s. Additionally, technological problems that resulted in frequent breakdowns of wind turbines and the much-publicized killing of birds of prey brought a negative image to the wind energy industry (Asmus 2001).

49. See McCright and Dunlap (2000, 2003); Newell (2000); and Lutzenhiser (2001).

50. See Bryner (2000); and Anderson (2002).

51. See "Wind Industry Decries Bias in Federal Tax Code," *Business Wire*, April 28, 1992.

52. See the Union of Concerned Scientists website http://www.ucsusa. org/clean_energy/solutions/big_picture_solutions/production-tax-credit-for.html (accessed December 2008).

53. Personal interview with a Sierra Club organizer, April 2008.

54. Personal interview with an Environment America organizer, March 2008.

55. Personal interview with a Sierra Club organizer, April 2008.

56. Personal interview with an Environment America organizer, March 2008.

57. "Comments of the American Wind Energy Association and the Union of Concerned Scientists on the Alternative Proposals Issued May 24, 1995, to Restructure California's Electric Services Industry and Reform Regulation," Docket No. 94-04-031, July 24, 1995.

58. As Wiser et al. (2007, 1) note, "RPS policies in Minnesota and Iowa predated the discussions in California, but these two renewable energy mandates were only later labeled as RPS policies."

59. For example, Rader and Norgaard (1996); Bernow, Dougherty, and Duckworth (1997); Haddad and Jefferiss (1999); and Wiser, Porter, and Clemmer (2000).

60. Michael Tennis, senior energy analyst, Union of Concerned Scientists, in the *Christian Science Monitor*, September 30, 1996.

61. See, for example, American Council for an Energy-Efficient Economy, Alliance to Save Energy, Natural Resources Defense Council, Tellus Institute, and Union of Concerned Scientists, *Energy Innovations: A Prosperous Path to a Clean Environment*, report no. E974 (Washington, D.C.: American Council for an Energy-Efficient Economy, 1997); Steven L. Clemmer, Alan Nogee, and Michael C. Brower, *A Powerful Opportunity: Making Renewable Electricity the Standard* (Cambridge, Mass.: Union of Concerned Scientists, 1999); and Union of Concerned Scientists, *Renewing Where We Live: What a National Renewable Electricity Standard Means for Your Region*, www.ucsusa.org/assets/documents/clean_energy/acfaok8cw.pdf (accessed May 2010).

62. See the Union of Concerned Scientists report http://www.ucsusa.org/clean_energy/solutions/renewable_energy_solutions/rps-campaign-national.html#benefits (accessed December 2008).

63. For example, the Edison Electric Institute, a trade association for America's investor-owned utilities, opposed a nationwide RPS, arguing that it would "raise consumers' electricity prices and create inequities among states" (http://www.eei.org/industry_issues/electricity_policy/federal_legislation/EEI_RPS.pdf). Moreover, the chairman of the House energy subcommittee charged with crafting energy and climate change legislation—Democrat Rick Boucher, from coal-rich southwest Virginia—consistently opposed a renewable electricity standard (http://public.cq.com/docs/gs/greensheets110-000002519747.html).

64. Personal interview with a Sierra Club organizer, April 2008.

65. Personal interview with an Environment America organizer, March 2008.

66. Personal interview with a Sierra Club organizer, April 2008.

67. See "NRDC, Other Environmental Groups Push for 33% California RPS by 2020," *Global Power Report*, August 28, 2008.

68. Personal interview with an Environmental Defense organizer, December 2007.

69. Ibid.

70. Personal interview with a representative of the Texas Renewable Energy Industries Association, November 2007.

71. Personal interview with a representative of the Texas Renewable Energy Industries Association, November 2007.

72. The World Future Council consists of a council of fifty personalities from around the globe who have already successfully promoted change on issues ranging from climate change to political, scientific, cultural, and economic justice. Its mission is "to inform and educate policymakers and opinion leaders about the challenges facing future generations while providing them with practical solutions. The WFC identifies and promotes successful

policies that can be implemented into legislation and policy measures" (http://www.worldfuturecouncil.org/about_us.html [accessed December 2008]).

73. See the World Future Council website http://www.worldfuturecouncil.org/climate_energy_campaign.html; emphasis in original.

74. World Future Council, "Spreading Good Policies: Power to the People!" http://www.worldfuturecouncil.org/pttp.html (accessed May 2010); emphasis in original.

75. See the Alliance for Renewable Energy website http://www.alliance-forrenewableenergy.org/press-release.html.

76. See the Wind-Works website http://www.wind-works.org/FeedLaws/USA/SCC%20FiT%20comments%20to%20CEC%2010102008.pdf (accessed November 2008).

77. See the Wind-Works website http://www.wind-works.org/index.htm (accessed December 2008).

78. See the Wind-Works website http://www.wind-works.org/FeedLaws/USA/Gipe%20Paul%20Comments%20CEC%20Feed-in%20Tariffs.pdf (accessed December 2008).

79. See the Wind-Works website http://www.wind-works.org/FeedLaws/USA/CaliforniaEnergyCommissionWeighsFeed-inTariff.html   (accessed December 2008).

80. See the Wind-Works website http://www.wind-works.org/FeedLaws/USA/TheCaliforniaFeed-inTariffActof2008.html (accessed December 2008).

81. See "US Rep. Inslee Introduces Renewable Energy Pricing Legislation," *Renewable Energy World*, June 27, 2008, http://www.renewableenergyworld.com/rea/news/story?id=52899 (accessed December 2008).

82. For more information about the U.S. fossil-fuel industry's antienvironmental campaigns, see Burton and Rampton (1997); Gelbspan (1998); Newell (2000); and McCright and Dunlap (2000).

83. Senator James Inhofe (R-Okla.), former chairman of the Senate Committee on Environment and Public Works, is one of the most vocal global warming skeptics in Congress. Here is an excerpt from one of his Senate Floor Statements, made in January 2005:

As I said on the Senate floor on July 28, 2003, "Much of the debate over global warming is predicated on fear, rather than science." I called the threat of catastrophic global warming the "greatest hoax ever perpetrated on the American people," a statement that, to put it mildly, was not viewed kindly by environmental extremists and their elitist organizations. I also pointed out, in a lengthy committee report, that those same environmental extremists exploit the issue for fundraising purposes, raking in millions of dollars, even using federal taxpayer dollars to finance their campaigns. For these groups, the issue of catastrophic global warming is not just a favored fundraising tool. In truth, it's more fundamental than that. Put simply, man-induced global warming is an article of religious faith. Therefore contending that its central tenets are flawed is, to them, heresy of the most despicable kind. Furthermore, scientists who challenge its tenets are attacked, sometimes personally, for blindly ignoring the so-called "scientific consensus" (http://inhofe.senate.gov/pressreleases/climateupdate.htm [accessed April 2009]).

84. Only 44 percent of the American public surveyed in August 2001 disapproved of President George W. Bush's decision to reject the Kyoto

Protocol. In comparison, 83 percent of the public in Great Britain and 87 percent of the public in Germany disapproved of this decision (Brechin 2003, 123).

85. For example, Marc Morano, a spokesman for Senator James Inhofe of Oklahoma, has expanded his campaign to convince legislators and the general public that global warming is nothing to worry about. According to a newspaper article, "Mr. Morano was for years a ceaseless purveyor of the dissenting view on climate change, sending out a blizzard of e-mail to journalists covering the issue. Now, with Congress debating legislation to curb carbon dioxide emissions, Mr. Morano is hoping to have an even greater impact. He has left his job with Mr. Inhofe to start his own Web site, ClimateDepot.com" (Leslie Kaufman, "Dissenter on Warming Expands His Campaign," New York Times, April 10, 2009, http://www.nytimes.com/2009/04/10/us/politics/10morano.html [accessed May 2010]).

86. The funding is intended to "support projects that draw on the innovations of DOE's national laboratories, universities, and the private sector to help improve reliability and overcome key technical challenges for the wind industry" (U.S. Department of Energy, http://apps1.eere.energy.gov/news/news_detail.cfm/news_id=12492 [accessed May 2009]).

87. See the North American Windpower website http://www.nawindpower.com/e107_plugins/content/content.php?content.5225 (accessed February 2010).

88. See Climate Change Solutions, "Renewable Portfolio Standard (RPS) & Other Incentives: Harmonization Opportunities in Canada," http://www.canbio.ca/documents/publications/Renewable_Harmonize_Final_Feb_16_2005.pdf (accessed June 2009); see also the Pembrina Institute website http://re.pembina.org/canada/policies (accessed June 2009).

89. Moreover, increasing demand for electricity and the desire to comply with Canada's Kyoto Protocol obligations have renewed government interest in nuclear energy, particularly in Ontario ("Ontario Considers Building a Nuclear Plant," http://www.electricityforum.com/news/jun04/ontnukes.html [accessed July 2009]).

90. In fact, the Campaign for Nuclear Phaseout logo, I Choose a Nuclear-Free Future for Canada, uses the image of wind turbines.

91. See Campaign for Nuclear Phaseout, "Phasing Out Nuclear Power in Canada: Toward Sustainable Electricity Futures," http://www.sierraclub.ca/national/programs/atmosphere-energy/nuclear-free/phasing-out-nuclear.pdf (accessed June 2009).

92. For information on Canada's coal reserves, see the Coal Association of Canada website http://www.coal.ca/content/index.php?option=com_content&task=section&id=9&Itemid=55 (accessed June 2009). For information on Canada's oil reserves, see the Energy Information Administration website http://www.eia.doe.gov/oiaf/ieo/pdf/table4.pdf (accessed June 2009). For information on Canada's uranium reserves, see the World Nuclear Association website http://www.world-nuclear.org/info/inf49.html (accessed June 2009).

93. See Climate Change Solutions, "Renewable Portfolio Standard."

94. See the Canadian Wind Energy Association website http://www.canwea.ca/pdf/Canada%20Current%20Installed%20Capacity_e.pdf (accessed October 2009).

95. Personal interview with Paul Gipe, November 2008.

96. Personal interview with Paul Gipe, November 2008.

97. The Ontario Sustainable Energy Association's main goals are to adopt a Green Energy Act in Ontario by 2010, to develop at least 500 MW of community power in Ontario by 2012, and to achieve 100 percent renewable energy powering Ontario by 2025 (Ontario Sustainable Energy Association, http://www.ontario-sea.org/ [accessed November 2008]).

98. Paul Gipe, Deborah Doncaster, and David MacLeod, "Powering Ontario Communities: Proposed Policy for Projects up to 10 MW, Study Outlining Policy Options to Encourage Small or Community-Owned Renewable Energy Generation in Ontario," http://www.wind-works.org/ FeedLaws/Canada/PoweringOntarioCommunities.pdf (accessed May 2010).

99. See the Ontario Sustainable Energy Association webpage http:// www.ontario-sea.org/Page.asp?PageID=122&ContentID=913&SiteNodeID= 205&BL_ExpandID= (accessed November 2008).

100. See Paul Gipe, "Ontario's Ruling Party Endorses Progressive Renewable Energy Policy—A First in North America," http://www. wind-works.org/FeedLaws/OntarioLiberalPartyEndorsesARTs.html (accessed November 2008).

Chapter 4

1. Brooklyn Brewery, http://www.brooklynbrewery.com/brewery/ (accessed November 2007).

2. Many other U.S. companies also purchase a significant percentage of their electricity from wind power; for example, banks (Wells Fargo, HSBC, and Citigroup), manufacturers (IBM, DuPont, and Johnson & Johnson), and retailers (Starbucks and Safeway).

3. See "Wal-Mart Commits to Wind Power," Greentech Media, November 20, 2008, http://www.greentechmedia.com/articles/read/wal-mart-commits-to-wind-power-5232/ (accessed May 2010).

4. Lori Bird, Leila Dagher, and Blair Swezey, *Green Power Marketing in the United States: A Status Report*, 10th ed., National Renewable Energy Laboratory, December 2007, http://www.eere.energy.gov/greenpower/ resources/pdfs/42502.pdf (accessed January 2008).

5. Rising Tide is a grassroots network of independent groups and individuals committed to taking action and building a movement against climate change. The network was formed in 2000 and is active mostly in the Netherlands, the United Kingdom, and other European countries. In 2006, a Rising Tide network was also formed in the United States; the network organized direct actions against new coal power plants ("Climate Crisis Energizes Radical Environmentalists," Associated Press Online, November 26, 2008).

6. For examples of recent studies that examine how social movements pressure organizations to change "from the outside," see Zald, Morill, and Rao (2005); King and Soule (2007); King (2008); Rao (2009); and Soule (2009).

7. Recent studies that examine how social movements contribute to organizational change "from the inside" include Moore (1999); Scully and Segal (2002); Raeburn (2004); and Strang and Jung (2005). For example, Moore shows that mediators initiate changes in organizations' actions because mediators occupy the middle ground between institutions and movements, and therefore they are in a good position to translate the claims of protesting groups into changes in practices and norms. Strang and Jung show that organization members who are also members or sympathizers of social movements can push for the adoption of new organizational practices either by launching an "orchestrated movement" (mobilizing for change from above), or by starting a grassroots movement (organizing for change from below).

8. See Granzin and Olsen (1991); Mainieri et al. (1997); Poortinga, Steg, and Vlek (2004); and Carlisle and Smith (2005).

9. For example, see Stets and Biga (2003).

10. Greenpeace Energy is one of the four independent green-power suppliers in Germany and sells green power to individuals, schools, churches, and public institutions. While most of its power comes from solar- and hydropower sources, it also purchases a significant amount of power from wind turbines in Austria and Denmark. Greenpeace Energy cannot purchase green power from German wind farms because the farms have to sell their electricity to utilities at rates set by the government for the first twenty years of operation (Vindenergi, http://www.vindenergi.dk/images/editor/pdf/germany.pdf [accessed May 2010]).

11. Personal interview with a California State University administrator, April 2007.

12. Personal interview with an Energy Action Coalition organizer, April 2007.

13. Personal interview with a New York University student activist, March 2007.

14. For example, an influential study of student activism on the issue of civil rights is McAdam (1988); for student activism against apartheid, see Soule (1997).

15. "Climate Change, One Light Bulb at a Time?" *Time.com*, November 8, 2007, http://www.time.com/time/health/article/0,8599,1682097,00.html.

16. "The Cool Club: In '60s Style, College Students Are Mobilizing to Work for Climate Change," *Boston Globe*, July 11, 2007, F1.

17. "Swarthmore College Makes Ambitious Commitment to Wind Power," Ascribe Newswire, February 28, 2007.

18. Personal interview with a cofounding member of the Energy Action Coalition, May 2007.

19. Jesse Jenkins, a member of the Cascade Climate Network, quoted in "Climate Change, One Light Bulb at a Time?" *Time.com*.

20. In 2000, Ozone Action was an organization without members but with a network of thousands of volunteers focusing on pollution that causes global warming or thins the earth's protective ozone layer. Greenpeace USA was an organization with a membership of about three hundred thousand and a budget of about $20 million a year, raised mostly by direct mail (John

Cushman Jr., "Arranging Environmental Groups' 'Corporate' Marriage," *New York Times*, August 7, 2000.

21. Western Resource Advocates (formerly known as the Land and Water Fund of the Rockies) was founded in 1989 and is "a non-profit environmental law and policy organization.... [whose] mission is to protect the West's land, air, and water.... to ensure a sustainable future for the West" (Western Resource Advocates, http://www.westernresourceadvocates.org/about/ [accessed May 2010]). It has offices in seven states (Colorado, Utah, Arizona, Nevada, New Mexico, Wyoming, and Idaho) and strategic programs in three areas: water, energy, and lands.

22. "A Summary of the University of Colorado Wind Power Campaign," http://ecenter.colorado.edu/energy/cu/renewables.html (accessed May 2008).

23. The Southern Alliance for Clean Energy was founded in 1981 under the name Tennessee Valley Energy Coalition. The Tennessee Valley Energy Coalition contributed to halting the Tennessee Valley Authority's plan to build seventeen nuclear power plants after the utility had completed five nuclear reactors and tallied a $28 billion debt (http://www.cleanenergy.org/).

24. See the Southern Alliance for Clean Energy website http://www.cleanenergy.org/ (accessed June 2008).

25. Personal interview with a Southern Alliance for Clean Energy organizer, April 2007.

26. Sierra Student Coalition, http://www.sierraclub.org/ssc/newsletter/ssc_Y-03-23.asp (accessed January 2008).

27. Energy Action Coalition, "Declaration of Independence from Dirty Energy," http://www.energyaction.net/documents/declaration.pdf (accessed December 2007).

28. Climate Challenge, http://www.climatechallenge.org/about (accessed January 2008).

29. Personal interview with a cofounding member of the Energy Action Coalition, May 2007.

30. See the Clean Air-Cool Planet website http://www.cleanair-coolplanet.org/champions/2004_sep.php (accessed January 2008).

31. Personal interview with a cofounding member of the Energy Action Coalition, May 2007.

32. Billy Parish originally became involved in climate change activism because of his conviction that "For my generation, global warming activism is a call to arms—because we are the ones who will have to deal with its growing consequences. And our efforts are at the same time a plea for accountability to the generation now in power; they have allowed the crisis to get to this point" (Clean Air-Cool Planet, http://www.cleanair-coolplanet.org/champions/2004_sep.php [accessed January 2008]).

33. Clean Air-Cool Planet was founded in 1999 by the Kendall Foundation when this foundation initiated a climate change program to focus specifically on steps to address accelerating greenhouse gas emissions (Kendall Foundation, http://www.kendall.org/about/). Clean Air-Cool Planet was founded with the intention of finding and promoting solutions to global warming in the Northeast (http://www.cleanair-coolplanet.org/about/).

34. In early 2008, the Campus Climate Challenge was supported by "The Bullitt Foundation, The Botwinick-Wolfensohn Family Foundation, Clear

the Air, The Energy Foundation, The Hull Family Foundation, The Kendall Foundation, The Kendeda Fund, Lynford Family Charitable Trust, Laird Norton Foundation, Merck Family Fund, New York Community Trust, Open Society Institute, The Overbrook Foundation, Rockefeller Brothers Fund, Surdna Foundation" (Energy Action Coalition, http://climatechallenge.org/about/supporters [accessed May 2010]).

35. The Environmental Center at the University of Colorado at Boulder was founded in 1970. It promoted the nation's first student-sponsored campus recycling program (in 1976), the first student-operated bus pass program (in 1991), and the first student-funded wind energy purchase (in 2000; "Presidents Climate Commitment Panel Discussion," http://www.colorado.edu/chancellor/speeches/climatecommitment.html [accessed June 2008]).

36. Personal interview with a Western Washington University official, February 2007.

37. Ibid.

38. According to one student activist: "Most students (and most citizens) have no clue where their energy comes from, which at UNC [University of North Carolina at Chapel Hill] is two-thirds coal and one-third nuclear power.... And, when most students learned about the dirty energy that the university currently uses, many of them were happy to pay a small fee to support cleaner renewable energy on campus" (Liz Veazy, a founding member of the referendum campaign, quoted in "Green Power to the People," *WireTap*, http://www.wiretapmag.org/stories/19895/ [accessed May 2010]).

39. Personal interview with a Green Arch student activist, February 2007.

40. Results from interviews show that the University of Central Oklahoma is the only case where the decision to buy wind power was made in the absence of a student campaign for clean energy.

41. Personal interview with a University of Central Oklahoma administrator, March 2007.

42. Personal interview with a Western Washington University student organizer, May 2007.

43. Ilse Kolbus, director of physical plant, University of California at Santa Cruz, http://www.epa.gov/greenpower/partners/partners/universityofcaliforniasantacruz.htm.

44. Personal interview with a Green Arch student activist, February 2007.

45. See http://www.aashe.org/about/about.php.

46. The American College & University Presidents Climate Commitment, http://www.presidentsclimatecommitment.org/html/faq.php#faq10 (accessed September 2007).

47. U.S. Department of Energy, The Green Power Network, "Large Purchasers of Green Power," http://www.eere.energy.gov/greenpower/buying/customers.shtml (accessed May 2010).

48. U.S. Environmental Protection Agency, Green Power Partnership, "Fortune 500 Partners," http://www.epa.gov/greenpower/toplists/fortune500.htm (accessed May 2010).

49. Personal interview with a Mohawk Fine Papers employee, May 2007.

50. Personal interview with an IBM employee, May 2007.

51. Personal interview with a World Resources Institute staff member, May 2007.

52. Personal interview with a Johnson & Johnson employee, May 2007.

53. Personal interview with a Whole Foods employee, May 2007.

54. Ibid.

55. World Resources Institute, http://www.wri.org/publication/safe-climate-sound-business-action-agenda (accessed June 2008).

56. World Resources Institute, http://www.wri.org/publication/safe-climate-sound-business-action-agenda (accessed May 2010).

57. The World Resources Institute, http://www.thegreenpowergroup.org/ (accessed June 2008).

58. See Edwards (2005, 2). Similarly, Stuart L. Hart argues that "the more we learn about the challenges of sustainability, the clearer it is that we are poised at the threshold of an historic moment in which many of the world's industries may be transformed" ("Beyond Greening: Strategies for a Sustainable World," *Harvard Business Review* (January–February, 1997), http://hbr.org/1997/01/beyond-greening/ar/1 [accessed May 2010]).

59. Sophia A. Muirhead et al., *Corporate Citizenship in the New Century: Accountability, Transparency, and Global Stakeholder Engagement* (New York: Conference Board, 2002).

60. See Reynolds, Donald. 2007. "Green Reporting in Business News Sections." http://www.businessjournalism.org/bizjournalism/Reynolds Center_GreenReport.pdf (accessed November 2007).

61. *Washington Post*, May 11, 2005, A17.

62. Personal interview with a World Resources Institute member, May 2007.

63. Personal interview with an IBM employee, May 2007.

64. Personal interview with a World Resources Institute member, May 2007.

65. Personal interview with a Johnson & Johnson employee, April 2007.

66. World Resources Institute, http://www.wri.org/project/green-power-markets.

67. See the Clean Air-Cool Planet case study at http://www.cleanair-coolplanet.org/documents/Timberland.pdf.

68. World Resources Institute, http://www.worldwildlife.org/climate/ item3799.html.

69. Center for Resource Solutions, http://www.resource-solutions.org/ who/history.htm (accessed July 2008).

70. Green-e, http://green-e.org/about_miss.shtml (accessed May 2010).

71. "Wells Fargo Commits to Largest-Ever Corporate Purchase of Renewable Energy in U.S.," https://www.wellsfargo.com/press/20061003_ GreenPower?year=2006 (accessed June 2008).

72. "Activists Demand Corporate Responsibility at San Francisco Investors Conference," http://www.commondreams.org/news2006/0921-04. htm (accessed June 2008).

73. Francesco Guerrera, "US Investor Group Unveils Climate Blacklist," *Financial Times*, February 13, 2007.

74. It is difficult to assess the exact influence of shareholder activism on corporate behavior. Companies usually deny that their actions are influenced by small groups of activists—either outside protesters or inside shareholders. However, anecdotal evidence suggests that these actions have effects that may range from greenwashing to taking meaningful steps such as increasing energy efficiency, using nonpolluting materials, and purchasing renewable energy. For anecdotes that illustrate the effect of environmental activism on the computer industry, see Smith, Sonnenfeld, and Pellow (2006).

75. The organization also calculates the environmental and economic benefits of pro-environment behavior; for example, by replacing three frequently used light bulbs with compact fluorescent bulbs, individuals could save three hundred pounds of carbon dioxide and sixty dollars per year; by moving the heater thermostat down two degrees in winter and up two degrees in summer, individuals could save two thousand pounds of carbon dioxide and ninety-eight dollars per year; and by insulating the water heater, people could save one thousand pounds of carbon dioxide and forty dollars per year (Stop Global Warming, http://www.stopglobalwarming.org/sgw_actionitems.asp [accessed December 2008]).

76. See http://www.ecotricity.co.uk/about/.

77. Personal interview with a Greenpeace Germany organizer, October 2008.

Chapter 5

1. Xcel Energy was formerly known as the Public Service Company of Colorado.

2. See the U.S. Department of Energy website http://apps3.eere.energy.gov/greenpower/markets/pricing.shtml?page=2&companyid=277 (accessed December 2008). See also the Xcel Energy website http://www.xcelenergy.com/Company/AboutUs/Pages/2007_Triple_Bottom_Line_Report.aspx (accessed December 2008).

3. The electricity sector includes companies that manufacture power plant equipment (General Electric, Siemens, etc.) and companies that develop and operate power plants (Xcel Energy, E.ON, etc.).

4. See, for example, "Energy Companies Make Wind Power a Top Investment," *International Herald Tribune*, June 4, 2007, http://www.iht.com/articles/2007/06/03/bloomberg/bxwind.php.

5. For example, in November 2005, Colorado Xcel Energy Windsource customers saved money relative to conventional energy costs, and the effective price of wind power remained below that of conventional power until April 2006. A similar situation occurred in July 2008 in Minnesota (http://apps3.eere.energy.gov/greenpower/markets/pricing.shtml?page=2&companyid=277). A spokesman for Vestas, the largest wind turbine manufacturer in the world, estimated in 2007 that an oil price of US$45 a barrel is "the threshold at which we're competitive with crude oil" ("Energy Companies Make Wind Power," *International Herald Tribune*).

6. For the purpose of this research, wind energy entrepreneurs are defined as founders of companies that manufacture wind turbines or develop and operate wind farms. Wind energy inventors formulate technical or financial innovations that contribute to wind turbine manufacturing or wind turbine development. Wind energy advocates support wind energy entrepreneurs and innovators by disseminating information and lobbying for the wind energy industry. Wind energy champions work for energy companies and push from inside to get them into wind turbine manufacturing or wind-farm development and operation. Obviously, in reality these theoretical constructs overlap significantly. For example, many wind energy entrepreneurs are also inventors or advocates, and many champions are also inventors and advocates.

7. Vestas remains the world's largest wind turbine manufacturer, although its market share decreased from 40 percent in 2004 to approximately 23 percent in 2008. According to the company website, "the company has installed more than 33,500 wind turbines in 62 countries and on five continents. Vestas installs an average of one wind turbine every five hours, twenty-four hours a day. Our wind turbines generate more than 50 million MWh [megawatt-hours] of energy per year, or enough electricity to supply millions of households" (http://www.vestas.com/en/about-vestas/history/2005-.aspx).

8. Tacke went bankrupt in 1997 and was bought by Enron (Pure Energy Professionals, "Industry Consolidation: Will It Increase or Decrease the Cost of WTGs?" presented at the European Wind Energy Association Conference, London, November 2004, www.2004ewec.info/files/24_1600_shanewood-roffe_01.pps [accessed May 2010]).

9. The influence of the Danish and German companies can be measured not only by the market share of companies headquartered in these countries but also by the role the companies played in the emergence of companies in other countries. For example, in 1994 Vestas, in collaboration with the Spanish group Gamesa and development company Sodena, established a joint venture company, Gamesa Eolica S.A., which is the largest Spanish wind turbine supplier today. In 1997, the German company Tacke was bought by Enron, and this Enron division later became part of General Electric.

10. See, for example, Heymann (1998); Karnøe, Kristensen, and Andersen (1999); Est (1999); Garud and Karnøe (2003); and Boon (2008).

11. Personal interview with Erik Grove-Nielsen, February 2009.

12. Ibid.

13. Ibid.

14. Ibid.

15. During World War II, the Danish company FL Smith erected several grid-connected turbines in Denmark (Erik Grove-Nielsen, Winds of Change, http://www.windsofchange.dk/WOC-77-81.php [accessed November 2008]).

16. See Grove-Nielsen, http://www.windsofchange.dk/WOC-77-81.php.

17. Grove-Nielsen, http://www.windsofchange.dk/WOC-77-81.php.

18. Ibid.

19. In 1984, Enercon started acquiring 7.5 meter blades from Coronet/AeroStar, the successor of Økær Vind Energi.

20. Personal interview with Henrik Stiesdal, November 2008.

21. Ibid.

22. Ibid.

23. Ibid.

24. Ibid.

25. Ibid.

26. See the Siemens website http://w1.siemens.com/innovation/en/innovators/energy/stiesdal.htm (accessed November 2008).

27. Personal interview with Henrik Stiesdal, November 2008.

28. Ibid.

29. Social movement scholars have argued that critical communities produce "interpretative packages" that identify problems and explain how they can be rectified, while also connecting social movement issues with dominant cultural themes. For example, see d'Anjou and Van Male (1998); and Rochon (1998).

30. See N. Meyer (2004, 26).

31. See N. Meyer (2004, 27–28).

32. Personal interview with Lars Albertsen, February 2009.

33. Ibid.

34. Ibid.

35. Ibid.

36. Personal interview with Preben Maegaard, March 2009.

37. Ibid.

38. Ibid.

39. Ibid.

40. See Preben Maegaard, "What Can We Learn from the Windmill History?" Tvindkraft, http://www.tvindkraft.dk/eng/Articel.asp?NewsID=20 (accessed April 2009).

41. Personal interview with Preben Maegaard, March 2009.

42. Ibid.

43. See K. Nielsen (2005, 109).

44. See Erik Grove-Nielsen, Winds of Change, http://www.windsofchange.dk/WOC-danturb.php (accessed April 2009).

45. Personal interview with Preben Maegaard, March 2009.

46. Ibid.

47. Ibid.

48. See Preben Maegaard, http://www.maegaard.net/maegaard_biography.html (accessed April 2009).

49. Maegaard, http://www.maegaard.net/maegaard_biography.html.

50. Personal interview with Preben Maegaard, March 2009.

51. Ibid.

52. Maegaard, http://www.maegaard.net/maegaard_biography.html.

53. See the Eurosolar website http://www.eurosolar.de/en/index.php?option=com_content&task=view&id=299&Itemid=91 (accessed May 2009).

54. See the Eurosolar website http://www.eurosolar.de/en/index.php?Itemid=27&id=221&option=com_content&task=view (accessed November 2008).

55. Mike Tidwell, director of the Chesapeake Climate Action Network, quoted in the New York Times, June 6, 2006, http://www.nytimes.com/2006/06/06/us/06wind.html.

56. Asmus (2001, 50).

57. Personal interview with Jim Dehlsen, March 2009.

58. Ibid.

59. See U.S. Department of Energy, "Wind Power Pioneer Interview: Jim Dehlsen, Clipper Windpower," http://www.windpoweringamerica.gov/filter_detail.asp?itemid=683 (accessed May 2010).

60. U.S. Department of Energy, "Wind Power Pioneer Interview."

61. See the Clipper Windpower website http://www.clipperwind.com/pdf/doe_award_091007.pdf (accessed March 2009).

62. See the Clipper Windpower website http://www.clipperwind.com/pr_121008.html (accessed March 2009).

63. Paul Gipe's role in the adoption of the first feed-in tariff (FIT) in North America is described in detail in chapter 3.

64. See Paul Gipe, Wind-Works, http://www.wind-works.org/bio.html (accessed May 2010).

65. Personal interview with Paul Gipe, November 2008.

66. Paul Gipe, Wind-Works, http://www.wind-works.org/giperesume.html (accessed October 2008).

67. Personal interview with Paul Gipe, November 2008.

68. See the Eurosolar website http://www.eurosolar.de/en/index.php?option=com_content&task=view&id=299&Itemid=91 (accessed May 2009).

69. Personal interview with James Lyons, March 2009.

70. See the New Energy New York website http://www.neny.org/ContentManager/index.cfm?Step=Display&ContentID=28 (accessed May 2009).

71. Personal interview with James Lyons, March 2009.

72. Ibid.

73. Ibid.

74. Ibid.

75. See Asmus (2001, 92).

76. See Hess (2007, 139).

77. For example, during the 1980s and early 1990s, a number of utilities in California were required by law to purchase a certain amount of power from renewable energy suppliers. In 1995, the Federal Energy Regulatory Commission struck down California's requirement—this led to the bankruptcy of wind power developers such as Kenetech Windpower (Hirsh 1999, 334).

78. For example, the London Array will eventually be the world's first 1 GW offshore wind farm. It is estimated that it will supply enough power for approximately 750,000 homes, or a third of Greater London homes. It will be built around twenty kilometers off the coasts of Kent and Essex. The project consortium partners have the following shareholdings: DONG Energy (50 percent), E.ON (30 percent), and Masdar (20 percent; http://www.londonarray.com/london-array-signs-final-phase-one/ [accessed May 2010]).

79. The nonutility or independent power companies, many of them developers of renewable energy projects, have also grown in some countries as a percentage of total capacity. In the United States, independent power companies grew from less than 3 percent of the market in the mid-1980s to over 8 percent in the mid-1990s, mostly because of the energy sector's deregulation (Hirsh 1999, 115).

80. See Hirsh (1999, 70). According to Hirsh, two other factors that challenged electricity companies' status as natural monopolies were the arrest of technological development and the energy crisis of the 1970s.

81. See Jane Kruse and Preben Maegaard, "Danish Wind Turbine History: An Authentic Story about How a Local Community Became Self-sufficient in Pollution Free Energy and Created a Source of Income for the Citizens," Folkecenter for Renewable Energy, http://www.folkecenter.net/gb/rd/wind-energy/history/ (accessed May 2009).

82. Obviously, not all cooperatives were formed by environmental activists or individuals who were motivated by an environmentalist ethic. In Denmark, for example, a number of communally owned wind energy parks established after 1985 were accused of being financial speculations (K. Nielsen 2005, 113).

83. See Paul Gipe, "Community Wind: The Third Way," Wind-Works, http://www.wind-works.org/articles/communitywindthethirdway.html [accessed May 2009].

84. N. Meyer (2004, 28).

85. See Dale Vince, Zero Carbonista, http://zerocarbonista.com/about-me/ (accessed March 2009).

86. Personal interview with Dale Vince, March 2009.

87. Ibid.

88. Ibid.

89. Personal interview with Stephen Tindale, December 2008.

90. In 1975 the Natural Resources Defense Council sued the Bonneville Power Administration, a federal agency that built hydroelectric dams in the Northwest. In 1977, the Natural Resources Defense Council, together with the Sierra Club, the Oregon Environmental Foundation, and the Northwest Fund for the Environment, published the Alternate Scenario, which focused on conservation measures to meet the region's growing demand for electricity (Hirsh 1999, 163).

91. The collaborative was created by the Conservation Law Foundation, a New England nonprofit environmental organization, and Connecticut Light and Power (Hirsh 1999, 211).

92. See Hirsh (1999, 220).

93. See Komor (2006, 133).

94. See Mayer, Blank, and Swezey (1999, 3).

95. Personal interview with John Nielson, Western Resource Advocates Energy Program director, February 2009.

96. Personal interview with Susan Innis, Western Resource Advocates Green-Power Marketing director, February 2009.

97. See Mayer, Blank, and Swezey (1999, 5).

98. Xcel conducted the review even though this was not required for a wind farm situated on private land. Regarding pricing, in the end "the parties agreed on a specific number and remained silent on the methodology used to reach the number." Finally, an agreement was made "to limit references to the environmental impacts of coal and instead to emphasize the positive benefits of renewables" (Mayer, Blank, and Swezey 1999, 4–6).

99. See Komor (2006, 135).

100. Personal interview with Fred Stoffell, Xcel Energy vice president of marketing, December 2008.

101. Personal interview with John Nielson, Western Resource Advocates Energy Program director, February 2009.

102. Personal interview with Susan Innis, Western Resource Advocates Green-Power Marketing director, February 2009.

103. See Komor (2006, 144).

104. See Komor (2006, 137).

105. See Komor (2006).

106. John Dunlop, American Wind Energy Association Great Plains representative, quoted in Paul Gipe, "The Great Wind Rush of 99," Wind-Works, http://www.wind-works.org/articles/99rush.html (accessed May 2010).

107. See Thomas Friedman, "Marching with a Mouse," New York Times, March 16, 2007, http://select.nytimes.com/2007/03/16/opinion/16friedman.html?sq=txu%20environmental%20defense&st=nyt&scp=3&pagewanted=print (accessed October 2008).

108. See Wade Goodwyn, "Critics Blast Texas Plans for New 'Dirty' Coal Plants," NPR, September 26, 2006, http://www.npr.org/templates/story/story.php?storyId=6110191 (accessed April 2009).

109. See "Ad Blitz Will Target TXU's Plant Plans," Austin American-Statesman, online edition, October 25, 2006, 8.

110. See the Environmental Defense website http://www.edf.org/documents/colin/TXU_FactSheet_Total_Packet.pdf (accessed April 2009).

111. For example, on February 21, 2007, activists from Rainforest Action Network staged protests outside the Boston offices of Merrill Lynch and tried to deliver a letter to its management ("Activists Stage Die-in at Merrill's Hub Office," Boston Herald, February 22, 2007, 23).

112. Personal interview with James Marston, Environmental Defense Texas Energy Program director, December 2007.

113. See Felicity Barringer and Andrew Ross Sorkin, "In Big Buyout, Utility to Limit New Coal Plants," New York Times, February 25, 2007, http://www.nytimes.com/2007/02/25/business/25coal.html (accessed December 2007).

114. Personal interview with James Marston, Environmental Defense Texas Energy Program director, December 2007.

115. See Barringer and Sorkin, "In Big Buyout."

116. For more information on the Center for Resource Solutions and the Green-e program, see Chapter 4.

117. See Friedman, "Marching with a Mouse."

118. See "Students Chain Selves to Duke," Raleigh News & Observer, November 16, 2007.

119. See "Eight Climate Protesters Arrested at U.S. Coal Plant," Reuters, April 1, 2008.

120. The complete list of coalition members included Southern Alliance for Clean Energy, North Carolina Interfaith Power & Light, Carolinas Clean Air Coalition, the Canary Coalition, Greenpeace USA, Rainforest Action Network, Mountain Voices Alliance, and Upper Watauga Riverkeeper (Stop Cliffside, http://stopcliffside.org/page.php?7 [accessed May 2009]).

121. "Hundreds March and 44 Arrested to Stop Cliffside Power Plant," Power Past Coal, April 21, 2009, http://www.powerpastcoal.org/article. php?id=152 (accessed May 2009).

122. See "A Green Coal Baron?" *New York Times*, June 22, 2008, http://www. nytimes.com/2008/06/22/magazine/22Rogers-t.html (accessed May 2009).

123. Founding members of the United States Climate Action Partnership include a number of major corporations: Alcoa, BP America, Caterpillar, Duke Energy, DuPont, FPL Group, General Electric, PG&E Corporation, and PNM Resources—and four NGOs: Environmental Defense, the Natural Resources Defense Council, the Pew Center on Global Climate Change, and the World Resources Institute (United States Climate Action Partnership, http://www.us-cap.org/ [accessed May 2009]).

124. Barringer and Sorkin, "In Big Buyout."

125. Personal interview with Gregory Efthimiou, Duke Energy communications manager, March 2009.

126. Barringer and Sorkin, "In Big Buyout."

127. Personal interview with Gregory Efthimiou, Duke Energy communications manager, March 2009.

128. Personal interview with Steven Stengel, NextEra Energy communications manager, March 2009.

129. Ibid.

130. Personal interview with Fred Stoffell, Xcel Energy vice president of marketing, December 2008.

131. Personal interview with Jim Dehlsen, March 2009.

132. See the Community Energy website http://www.newwindenergy. com/about-us/about-us-summary/ (accessed March 2009).

133. See Jeff Gelles, "The Electricity of Innovation," *Philadelphia Inquirer*, October 17, 2005, C1.

134. See "Spain's Iberdrola to Acquire Community Energy, Inc," *Renewable Energy Today*, May 4, 2006, http://findarticles.com/p/articles/ mi_m0OXD/is_2006_May_4/ai_n16347327/ (accessed March 2009).

135. According to one estimate, the Alliance would ultimately spend well over $ 10 million trying to keep the Cape Wind project from being permitted (Williams and Whitcomb 2007: 95).

136. For more information on the Alliance's campaign, see Williams and Whitcomb (2007).

137. Personal interview with Mark Rodgers, Communications Director for Energy Management Inc., March 2008.

138. Ibid.

139. See the "List of Supporters" from Cape Wind's website, at: http:// www.capewind.org/article47.htm (accessed June 2008)

140. See Williams and Whitcomb (2007: 228).

Conclusion

1. See the United Nations Commission of Sustainable Development website http://www.un.org/esa/sustdev/csd/csd14/bgrounder_energyforall. pdf (accessed May 2009).

2. See "New Research Reveals Most Americans Don't Know Where Their Electricity Comes From," *EnviroMedia*, http://www.enviromedia.com/news-item.php?id=197 (accessed May 2010).

3. U.S. Energy Information Administration, "International Energy Outlook 2008," http://www.eia.doe.gov/oiaf/ieo/electricity.html (accessed May 2009).

4. World Wind Energy Association, "World Wind Energy Report 2008," http://www.wwindea.org/home/index.php?option=com_content&task=view&id=226&Itemid=43 (accessed May 2010).

5. U.S. Energy Information Administration, "International Energy Outlook 2008."

6. See REN21 (Renewable Energy Policy Network for the 21st Century), "Renewables Global Status Report: 2009 Update," http://www.ren21.net/pdf/RE_GSR_2009_update.pdf (accessed May 2009).

7. The Climate Action Network (CAN) is an umbrella organization that includes over 450 environmental NGOs with seven regional offices "working to promote government and individual action to limit human-induced climate change to ecologically sustainable levels" (Climate Action Network, http://www.climatenetwork.org/about-can [accessed March 2009]). For analyses of the role of NGOs in climate change negotiations, see Newell (2000); and Fisher (2004).

8. See Global Wind Energy Council, "Global Wind Energy Outlook 2008," http://www.gwec.net/fileadmin/documents/Publications/GWEO_2008_final.pdf (accessed May 2010), 54.

9. Global Wind Energy Council, "Global Wind Energy Outlook 2008."

10. For an analysis of types of environmental discourse, see Dryzek (1997).

11. See Amanda Little, "RFK Jr. and Other Prominent Enviros Face Off over Cape Cod Wind Farm," http://www.grist.org/article/capecod/ (accessed May 2010).

12. Here is one example of a headline that describes the environmental community as divided over wind farms: "Activists Are Split on a Proposed Wind Project Off Cape Cod" (*Grist*, http://www.grist.org/article/griscom-windmill/ [accessed April 2009]).

13. See Natural Resources Defense Council, "Cape Wind Can Start Leading the Way for Offshore Wind Power," http://switchboard.nrdc.org/blogs/fbeinecke/cape_wind_can_start_leading_th.html (accessed May 2009).

14. The role of influential individuals with ties to the oil and coal industry in the anti–Cape Wind campaign is documented in Williams and Whitcomb (2007).

15. See "Audubon Society Backs Controversial Wind Farm," Associated Press, March 29, 2006, http://www.msnbc.msn.com/id/12066651/from/ET/ (accessed April 2009).

16. See the Audubon Society website http://www.audubon.org/campaign/windPowerQA.html (accessed April 2009).

17. See the Royal Society for the Protection of Birds website http://www.rspb.org.uk/ourwork/policy/windfarms/index.asp (accessed April 2009).

18. It is important to note that the Danish grid is heavily interconnected with the European electrical grid, and Denmark has solved grid-management

problems by exporting almost half of its wind power to Norway. However, for the future Denmark has active plans to increase the percentage of power generated by wind to over 50 percent of total capacity ("Analysis of Wind Power in the Danish Electricity Supply in 2005 and 2006," Techconsult, http://www.wind-watch.org/documents/wp-content/uploads/dk-analysis-wind.pdf [accessed May 2009]).

19. See Sinclair Merz, "Growth Scenarios for UK Renewables Generation and Implications for Future Developments and Operation of Electricity Networks," http://www.berr.gov.uk/files/file46772.pdf (accessed May 2009).

20. This can be achieved because HVDC has the ability to efficiently shift power from windy areas to non-windy areas; the intermittency problem could be solved by geographic dispersion to decouple weather-system effects (Gregor Czisch and Gregor Giebel, "Realisable Scenarios for a Future Electricity Supply Based 100% on Renewable Energies," http://www.risoe.dk/rispubl/reports/ris-r-1608_186-195.pdf [accessed May 2010]).

21. The main challenges are increased corrosion due to salty air, and higher maintenance and operation costs; for example, the first deep-sea wind farm, the Alpha Ventus project located in Germany, is estimated to cost nearly three times as much as a similar installation on land ("A Green Revolution off Germany's Coast," *Spiegel Online International*, July 24, 2008, http://www.spiegel.de/international/germany/0,1518,567622,00.html [accessed May 2009]). for information about existing offshore wind turbines, see the German Wind Energy Association website http://www.wind-energy-market.com/index.php?id=4&no_cache=1&L=0 (accessed May 2009). For information about offshore systems, see the Sway website http://sway.no/index.php?id=16 (accessed May 2009).

22. See European Wind Energy Association, "Wind at Work: Wind Energy and Job Creation in the EU," http://www.ewea.org/fileadmin/ewea_documents/documents/publications/Wind_at_work_FINAL.pdf (accessed May 2009).

23. European Wind Energy Association, "Wind at Work."

24. See Lester Brown, "Farmers Are Reaping Rewards from Wind Energy," *Grist*, September 21, 2000, http://www.grist.org/article/brown-something/ (accessed May 2009).

25. Global Wind Energy Council, "Global Wind Energy Outlook 2008."

26. See Keith Bradsher, "The Ascent of Wind Power," *New York Times*, September 28, 2006, http://www.nytimes.com/2006/09/28/business/world-business/28wind.html (accessed September 2008).

27. European Wind Energy Association, "Wind at Work."

28. See the Global Energy Network Institute website http://www.geni.org/globalenergy/library/national_energy_grid/norway/index.shtml (accessed May 2009).

29. This does not imply that efforts concentrating on raising awareness are not important; in fact, educating the public and raising awareness of an injustice or social problem is a basic activity for most social movement organizations.

30. See Janet Sawin, "Another Sunny Year for Solar Power," Worldwatch Institute, May 8, 2008, http://www.worldwatch.org/node/5449 (accessed May 2009).

31. One article argues that "Germany, as a technology leader, reached investment levels of around €700 million in 2006 with 400 companies involved in segments of the sector and 100 companies offering the whole value chain. Export of biogas-related technologies and services is between 10 and 15 percent" ("Market Study Tracks Global Boom in Biogas, Germany Technology Leader," http://news.mongabay.com/bioenergy/2007/07/market-study-tracks-global-boom-in.html [accessed August 2009]).

32. The best-known organization is Plug In America, whose mission is to "accelerate the shift to plug-in vehicles powered by clean, affordable, domestic electricity to reduce our nation's dependence on petroleum and improve the global environment" (Plug In America, http://www.plugina-merica.org/about-us/about-us.html [accessed August 2009]).

33. On the importance of environmental activists in the early days of the renewable energy industries, see Hess (2007, 139).

34. Wind energy companies have already reported an acute shortage of workers, particularly engineers and operation and management specialists. To remedy this, European wind energy professionals have suggested creating a European wind energy training center, with correspondents in various countries and local partnerships with universities. See the European Wind Energy Association's website, http://www.ewea.org/index.php?id=1638 (accessed May 2010).

35. The Association for the Study of Peak Oil and Gas predicted in its January 2008 newsletter that the peak in all oil sources (including nonconventional) would occur in 2010 (http://www.aspo-ireland.org/contentFiles/newsletterPDFs/newsletter85_200801.pdf [accessed May 2009]). The International Energy Agency predicted a plateau in the global production of oil in 2020 and a peak by 2030 (George Monbiot, "When Will the Oil Run Out?" *Guardian*, December 15, 2008, http://www.guardian.co.uk/business/2008/dec/15/oil-peak-energy-iea [accessed May 2009]).

36. See the National Energy Technology Laboratory website http://www.netl.doe.gov/publications/others/pdf/Oil_Peaking_NETL.pdf (accessed May 2009).

37. For example, Deffeyes (2003); Heinberg (2005); Roberts (2005); Tertzakian (2006); and Friedman (2008).

38. See PickensPlan, http://www.pickensplan.com/theplan/ (accessed May 2009).

39. See "Pickens Calls Off Plans for Vast Texas Wind Farm," *Washington Post*, July 8, 2009, http://www.washingtonpost.com/wp-dyn/content/article/2009/07/07/AR2009070702455.html (accessed August 2009).

40. See "Titan Wind Project to Produce 5,050 MW," http://www.renewableenergyworld.com/rea/news/article/2008/08/titan-wind-project-to-produce-5050-mw-53232 (accessed May 2010).

41. See "Floating Turbines May Join Norway's Offshore Rigs," Reuters, April 29, 2008, http://uk.reuters.com/article/environmentNews/idUKL2589097520080429?pageNumber=3&virtualBrandChannel=0 (accessed May 2009).

42. See Josh Wolfe, "Nuclear Renaissance," *Forbes*, November 14, 2007, http://www.forbes.com/2007/11/14/emergingtech-nuclear-renaissance-pf-guru-in_jw_1114advisersoapbox_inl.html (accessed May 2009).

43. See the World Nuclear Association website http://www.world-nuclear.org/info/inf41.html#licence (accessed May 2009).

44. See "Obama Announces First New Nuclear Power Plant in US for 30 Years," http://www.france24.com/en/20100216-barack-obama-announces-plans-new-nuclear-power-plants-usa (accessed February 2010).

45. "New Nuclear Plants Get Go-Ahead," *BBC News*, January 10, 2008, http://news.bbc.co.uk/2/hi/uk_news/politics/7179579.stm (accessed May 2009).

46. "New Nuclear Plants Get Go-Ahead," *BBC News*.

47. See the American Coalition for Clean Coal Electricity website http://www.cleancoalusa.org/docs/issue/ (accessed February 2010).

48. See Jeff Biggers, "'Clean' Coal? Don't Try to Shovel That," *Washington Post*, March 2, 2008, http://www.washingtonpost.com/wp-dyn/content/article/2008/02/29/AR2008022903390.html.

49. "Old King Coal: Why the World Needs a Coal Power Moratorium," Worldchanging, July 8, 2007, http://www.worldchanging.com/archives/006997.html (accessed May 2009).

50. To find out more about the "Schwarzenegger clause," see Natural Resources Defense Council, "California Takes On Power Plant Emissions: SB 1368 Sets Groundbreaking Greenhouse Gas Performance Standard," http://www.solutionsforglobalwarming.org/docs/SB1368_FS_FINAL.pdf(accessed May 2009).

51. See Opinion Research Corporation and Citizens Lead for Energy Action Now (CLEAN), "A Post Fossil-Fuel America: Are Americans Ready to Make the Shift?" October 18, 2007, http://v1.apebble.com/static/101807_CLEAN_survey_report.pdf (accessed May 2009).

52. See SourceWatch, "Coal Plants Cancelled in 2007," http://www.sourcewatch.org/index.php?title=Coal_plants_cancelled_in_2007 (accessed May 2009).

53. See European Parliament, "Equipping Power Plants to Store CO2 Underground," July 10, 2008, http://www.europarl.europa.eu/sides/getDoc.do?language=EN&type=IM-PRESS&reference=20081006IPR38802 (accessed May 2010).

54. See "Chairman Waxman Introduces H.R. 5575, the 'Moratorium on Uncontrolled Power Plants Act,'" http://oversight.house.gov/story.asp?ID=1797 (accessed May 2009).

55. The environmental groups' statement argued that "despite the best efforts of Chairman Waxman, the decision-making process was co-opted by oil and coal lobbyists determined to sustain our addiction to dirty fossil fuels. The resulting bill reflects the triumph of politics over science, and the triumph of industry influence over public interest" (John Broder, "Climate Bill Clears Hurdle, but Others Remain," *New York Times*, May 21, 2009, http://www.nytimes.com/2009/05/22/us/politics/22climate.html?_r=1&th&emc=th [accessed May 2009]).

References

Ackermann, Thomas, ed. 2005. *Wind Power in Power Systems*. Chichester, U.K.: John Wiley.

Aldrich, Howard, and Marlene Fiol. 1994. "Fools Rush In? The Institutional Context of Industry Creation." *Academy of Management Review* 19:645–70.

Almanzar, Nelson Pichardo, Heather Sullivan-Catlin, and Glenn Deane. 1998. "Is the Political Personal? Everyday Behaviors as Forms of Environmental Participation." *Mobilization* 3:185–205.

Amenta, Edwin. 2006. *When Movements Matter: The Townsend Plan and the Rise of Social Security*. Princeton, N.J.: Princeton University Press.

Amenta, Edwin, and Neal Caren. 2004. "The Legislative, Organizational, and Beneficiary Consequences of State-Oriented Challengers." In *The Blackwell Companion to Social Movements*, edited by David Snow, Sarah Soule, and Hanspeter Kriesi, 461–89. Malden, Mass.: Blackwell Publishing.

Amenta, Edwin, and Michael Young. 1999. "Making an Impact: Conceptual and Methodological Implications of the Collective Goods Criterion." In *How Social Movements Matter*, edited by Marco Giugni, Doug McAdam, and Charles Tilly, 22–41. Minneapolis: University of Minnesota Press.

Andersen, Svein, and Atle Midttun. 1994. "Environmental Opposition in Norwegian Energy Policy: Structural Determinants and Strategic Mobilization." In *States and Anti-nuclear Movements*, edited by Helena Flam, 232–63. Edinburgh: Edinburgh University Press.

Anderson, Kai. 2002. "The Climate Policy Debate in the US Congress." In *Climate Change Policy: A Survey*, edited by Stephen Schneider, Armin Rosencranz, and John Niles, 235–51. Washington, D.C.: Island Press.

Andrews, Kenneth. 2001. "Social Movements and Policy Implementation: The Mississippi Civil Rights Movement and the War on Poverty, 1965–71." *American Sociological Review* 66:71–95.

Archer, Cristina L., and Mark Z. Jacobson. 2005. "Evaluation of Global Wind Power." *Journal of Geophysical Research* 110 (June 30).

Asmus, Peter. 2001. *Reaping the Wind: How Mechanical Wizards, Visionaries, and Profiteers Helped Shape Our Energy Future.* Washington, D.C.: Island Press.

Babin Ronald. 1985. *The Nuclear Power Game.* Black Rose Books.

Baum, Joel, and Christine Oliver. 1992. "Institutional Embeddedness and the Dynamics of Organizational Populations." *American Sociological Review* 57:540–59.

Beisheim, Marianne. 2005. "Politics from Above or Below? Climate Politics in Germany and Great Britain." In *Globalizing Interests: Pressure Groups and Denationalization*, edited by Michael Zurn and Gregor Walter, 187–256. Albany: State University of New York Press.

Bernow, Steve, William Dougherty, and Max Duckworth. 1997. "Can We Afford a Renewables Portfolio Standard?" *Electricity Journal* (May).

Beuermann, Christiane, and Jill Jäger. 1996. "Climate Change politics in Germany: How Long Will Any Double Dividend Last?" In *Politics of Climate Change: A European Perspective*, edited by Tim O'Riordan and Jill Jäger, 186–227. London: Routledge.

Bhadra, Sailendra N., Debaprasad Kastha, and Soumitro Banerjee. 2005. *Wind Electrical Systems.* New York: Oxford University Press.

Blanco, Maria Isabel, and Glória Rodrigues. 2009. "Direct Employment in the Wind Energy Sector: An EU study." *Energy Policy* 37 (8): 2847–57.

Boon, Martin. 2008. "Why Did Danish Entrepreneurs Take the Lead in the Wind Turbine Industry and Not the Dutch? A Study on the Interaction between Evolution and Strategy of Two Communities in an Emerging Industry." Master's thesis.

Bosso, Christopher. 2000. "Environmental Groups and the New Political Landscape." In *Environmental Policy*, edited by Norman Vig and Michael Kraft, 4th ed. Washington, D.C.: CQ Press, 55–76.

———. 2005. *Environment, Inc.: From Grassroots to Beltway.* Lawrence: University Press of Kansas.

Boykoff Maxwell and Jules Boykoff. 2004. "Balance as Bias: Global Warming and the U.S. Prestige Press." *Global Environmental Change*, 14: 125–136.

Brand, Karl-Werner. 1999. "Dialectics of Institutionalization: The Transformation of the Environmental Movement in Germany." *Environmental Politics* 8 (1): 35–58.

Brechin, Steven. 2003. "Comparative Public Opinion and Knowledge on Global Climatic Change and the Kyoto Protocol: The U.S. versus the World?" *International Journal of Sociology and Public Policy* 23:106–34.

Brulle, Robert J. 2000. *Agency, Democracy, and Nature: The U. S. Environmental Movement from a Critical Theory Perspective.* Cambridge, Mass.: MIT Press.

Bryner, Gary. 2000. "Congress and the Politics of Climate Change." In *Climate Change and American Foreign Policy*, edited by Paul Harris, 111–30. New York: Saint Martin's Press.

Burstein, Paul. 1999. "Social Movements and Public Policy." In *How Social Movements Matter*, edited by Marco Giugni, Doug McAdam, and Charles Tilly, 3–21. Minneapolis: University of Minnesota Press.

Burstein, Paul, and April Linton. 2002. "The Impact of Political Parties, Interest Groups, and Social Movement Organizations on Public Policy: Some Recent Evidence and Theoretical Concerns." *Social Forces* 81:380–408.

Burton, Bob, and Sheldon Rampton. 1997. "Thinking Globally, Acting Vocally: The International Conspiracy to Overheat the Earth." http://www.prwatch.org/prwissues/1997Q4/warming.html (accessed June 2006).

Buttel, Frederick, and Peter Taylor. 1994. "Environmental Sociology and Global Environmental Change: A Critical Assessment." In *Social Theory and the Global Environment*, edited by Michael Redclift and Ted Benton, 228–55. London: Routledge.

Carlisle, Juliet, and Eric Smith. 2005. "Postmaterialism vs. Egalitarianism as Predictors of Energy-Related Attitudes." *Environmental Politics* 14:527–40.

Chandler, Alfred Jr. 2001. *Inventing the Electronic Century: The Epic Story of the Consumer Electronics and Computer Industries*. New York: Free Press.

Chernow, Ron. 1998. *Titan: The Life of John D. Rockefeller*. New York: Vintage Books.

Coenen, Reinhard. 1999. "Science and the Policy Process in Germany." In *Climate Change Policy in Germany and the United States: Conference Proceedings, Berlin, June 15–18, 1997*, edited by the Deutsch-Amerikanisches Akademisches Konzil [DAAK; German-American Academic Council Foundation], 224–32. Bonn: DAAK.

Connor, Peter. 2005. "The UK Renewables Obligation." In *Switching to Renewable Power: A Framework for the 21st Century*, edited by Volkmar Lauber. London: Earthscan.

Cress, Daniel M., and David A. Snow. 2000. "The Outcomes of Homeless Mobilization: The Influence of Organization, Disruption, Political Mediation, and Framing." *American Journal of Sociology* 105:1063–1104.

Dalton, Russell. 2005. "The Greening of the Globe? Cross-national Levels of Environmental Group Membership." *Environmental Politics* 14:441–59.

d'Anjou, Leo, and John Van Male. 1998. "Between Old and New: Social Movements and Cultural Change." *Mobilization* 3:207–26.

Davis, Gerard, Kristina Diekmann, and Catherine Tinsley. 1994. "The Deinstitutionalization of the Conglomerate Firms in the 1980s." *American Sociological Review* 59:547–70.

Deffeyes, Kenneth. 2003. *Hubbert's Peak: The Impending World Oil Shortage*. Princeton, N.J.: Princeton University Press.

Derksen, Linda, and John Gartrell. 1993. "The Social Context of Recycling." *American Sociological Review* 58:434–42.

Diani, Mario. 1995. *Green Networks: A Structural Analysis of the Italian Environmental Movement*. Edinburgh: Edinburgh University Press.

———. 2004. "Networks and Participation." In *The Blackwell Companion to Social Movements*, edited by David Snow, Sarah Soule, and Hanspeter Kriesi, 339–59. Malden, Mass.: Blackwell Publishing.

Dietz, Thomas, and Linda Kalof. 1992. "Environmentalism among Nation-States." *Social Indicators Research* 26:353–66.

DiMaggio, Paul J. 1991. "Constructing an Organizational Field as a Professional Project: U.S. Art Museums, 1920–1940." In *The New Institutionalism in Organizational Analysis*, edited by Walter W. Powell and Paul J. DiMaggio, 267–92. Chicago: University of Chicago Press.

———, ed. 2001. *The Twenty-First-Century Firm: Changing Economic Organization in International Perspective*. Princeton, N.J.: Princeton University Press.

DiMaggio, Paul J., and Walter W. Powell. 1983. "The Iron Cage Revisited: Institutional Isomorphism and Collective Rationality in Organizational Fields." *American Sociological Review* 48:147–60.

Dispensa Jaclyn Marisa and Robert Brulle. 2003. "Media's Social Construction of Environmental Issues: Focus on Global Warming –a Comparative Study." *International Journal of Sociology and Social Policy*, 23: 74–105.

Dobbin, Frank, and Timothy Dowd. 1997. "How Policy Shapes Competition: Early Railroad Foundings in Massachusetts." *Administrative Science Quarterly* 42:501–29.

———. 2000. "The Market that Antitrust Built: Public Policy, Private Coercion, and Railroad Acquisitions, 1825 to 1922." *American Sociological Review* 65:631–57.

Douthwaite, Boru. 2002. *Enabling Innovation: A Practical Guide to Understanding and Fostering Technological Change*. London: Zed Books.

Dowie, Mark. 1995. *Losing Ground: American Environmentalism at the Close of the Twentieth Century*. Cambridge, Mass.: MIT Press.

Dryzek, John S. 1997. *The Politics of the Earth: Environmental Discourses*. Oxford: Oxford University Press.

Earl, Jennifer. 2004. "The Cultural Consequences of Social Movements." In *How Social Movements Matter*, edited by Marco Giugni, Doug McAdam, and Charles Tilly, 508–30. Minneapolis: University of Minnesota Press.

Eckersley, Robyn. 2004. *The Green State. Rethinking Democracy and Sovereignty*. Cambridge, Mass: MIT Press.

Edelman, Lauren. 1990. "Legal Environments and Organizational Governance: The Expansion of Due Process in the American Workplace." *American Journal of Sociology* 95:1401–40.

Edwards Andres. 2005. *The Sustainability Revolution. Portrait of a Paradigm Shift*. New Society Publishers.

Erickson, Wallace, Gregory Johnson, and David Young. 2005. "A Summary and Comparison of Bird Mortality from Anthropogenic Causes with an Emphasis on Collisions." In *Bird Conservation Implementation and Integration in the Americas: Proceedings of the Third International Partners in Flight Conference*, 2002 March 20–24, Asilomar, California, Vol. 2 Gen. Tech. Rep. PSW-GTR-191, edited by C. John Ralph and Terrell D. Rich, 1029–42. Albany, Calif.: U.S. Department of Agriculture, Forest Service, Pacific Southwest Research Station.

Est, Quirinus Cornelis van. 1999. *Winds of Change: A Comparative Study of the Politics of Wind Energy Innovation in California and Denmark*. Utrecht: International Books.

Esty, Daniel C., Marc Levy, Tanja Srebotnjak, and Alexander de Sherbinin. 2005. *Environmental Sustainability Index: Benchmarking National Environmental Stewardship*. New Haven, Conn.: Yale Center for Environmental Law and Policy.

European Wind Energy Association. 2003. "Focus on Public Opinion: A Summary of Opinion Surveys on Wind Power." http://www.ewea.org/fileadmin/ewea_documents/documents/publications/WD/WD22vi_public.pdf (accessed May 2010).

Eyerman, Ron, and Andrew Jamison. 1991. *Social Movements: A Cognitive Approach*. University Park: Pennsylvania State University Press.

Fillieule Olivier. 2003. "France", in *Environmental Protest in Western Europe*, edited by Christopher Rootes, 59–79. Oxford: Oxford University Press.

Fisher, Dana. 2004. *National Governance and the Global Climate Change Regime*. Lanham, Md.: Rowman and Littlefield.

———. 2006. "Bringing the Material Back In: Understanding the U.S. Position on Climate Change." *Sociological Forum* 21:467–94.

Flam, Helena, and Andrew Jamison. 1994. "The Swedish Confrontation over Nuclear Energy: A Case of a Timid Anti-nuclear Opposition." In *States and Anti-nuclear Movements*, edited by Helena Flam, 163–201. Edinburgh: Edinburgh University Press.

Fligstein, Neil. 1990. *The Transformation of Corporate Control*. Cambridge, Mass.: Harvard University Press.

———. 1996. "Markets as Politics: A Political-Cultural Approach to Market Institutions." *American Sociological Review* 61:656–73.

Forbes, Linda C., and John M. Jermier. 2002. "The Institutionalization of Voluntary Organizational Greening and the Ideals of Environmentalism: Lessons about Official Culture from Symbolic Organization Theory." In *Organizations, Policy, and the Natural Environment: Institutional and Strategic Perspectives*, edited by Andrew J. Hoffman and Marc J. Ventresca, 194–213. Palo Alto, Calif.: Stanford University Press.

Frank, David John, Ann Hironaka, and Evan Schofer. 2000. "The Nation State and the Natural Environment over the Twentieth Century." *American Sociological Review* 65:96–116.

Freeland, Robert. 2001. *The Struggle for Control of the Modern Corporation: Organizational Change at General Motors, 1924–1970*. Cambridge: Cambridge University Press.

Freudenburg, William R., Scott Frickel, and Robert Gramling. 1995. "Beyond the Nature/Society Divide: Learning to Think about a Mountain." *Sociological Forum* 10:361–92.

Freudenburg, William R., and Robert Gamling. 1994. *Oil in Troubled Waters: Perceptions, Politics, and the Battle over Offshore Drilling*. Albany: State University of New York Press.

Friedman, Thomas. 2008. *Hot, Flat, and Crowded: Why We Need a Green Revolution, and How It Can Renew America*. New York: Farrar, Straus and Giroux.

Garud, Raghu, and Peter Karnøe. 2003. "Bricolage versus Breakthrough: Distributed and Embedded Agency in Technology Entrepreneurship." *Research Policy* 32:277–300.

Gelbspan, Ross. 1998. *The Heat Is On: The Climate Crisis, the Cover-Up, the Prescription*. Reading, Mass.: Perseus Publishing.

Gerhards, Jurgen, and Dieter Rucht. 1992. "Mesomobilization: Organizing and Framing in Two Protest Campaigns in West Germany." *American Journal of Sociology* 98:555–95.

Gipe, Paul. 1995. *Wind Energy Comes of Age*. New York: Wiley.

———. 2004. *Wind Power: Renewable Energy for Home, Farm, and Business*. White River Junction, Vt.: Chelsea Green Publishing.

———. 2006. "Renewable Energy Policy Mechanisms." Unpublished paper. http://www.wind-works.org/FeedLaws/RenewableEnergyPolicyMechanisms byPaulGipe.pdf.

Giugni, Marco. 2004. *Social Protest and Policy Change: Ecology, Antinuclear, and Peace Movements in Comparative Perspective*. Lanham, Md.: Rowman and Littlefield.

———. 2007. "Useless Protest? A Time-Series Analysis of the Policy Outcomes of Ecology, Anti-nuclear, and Peace Movements in the United States, 1977–1995." *Mobilization* 12:53–77.

Giugni, Marco, Doug McAdam, and Charles Tilly, eds. 1999. *How Social Movements Matter*. Minneapolis: University of Minnesota Press.

Goodell, Jeff. 2006. *Big Coal: The Dirty Secret behind America's Energy Future*. Boston: Houghton Mifflin.

Granovetter, Mark. 1985. "Economic Action, Social Structure, and Embeddedness." *American Journal of Sociology* 91:481–510.

Granzin, Kent, and Janeen Olsen. 1991. "Characterizing Participants in Activities Protecting the Environment: A Focus on Donating, Recycling, and Conservation Behaviors." *Journal of Public Policy & Marketing* 10:1–27.

Griffin, James, and Steven Puller. 2005. *Electricity Deregulation: Choices and Challenges*. Chicago: University of Chicago Press.

Haddad, Brent, and Paul Jefferiss. 1999. "Forging Consensus on National Renewables Policy: The Renewables Portfolio Standard and the National Public Benefits Trust Fund." *Electricity Journal* (March).

Hannigan, John. 1995. *Environmental Sociology: A Social Constructionist Perspective*. London: Routledge.

Hansen, Anca. 2005. "Generators and Power Electronics for Wind Turbines." In *Wind Power in Power Systems*, edited by Thomas Ackermann, 53–77. John Wiley and Sons, Ltd.

Harris, C. D. 1954. "The Market as a Factor in the Location of Industry in the United States." *Annals of the Association of American Geographers* 44:315–48.

Harrison, Robert Pogue. 2002. "Hic Jacet." In *Landscape and Power*, edited by W. J. T. Mitchell, 2nd ed., 349–64. Chicago: University of Chicago Press.

Hatch, Michael. 1986. *Politics and Nuclear Power: Energy Policy in Western Europe*. Lexington: University Press of Kentucky.

Haveman, Heather, and Hayagreeva Rao. 1997. "Structuring a Theory of Moral Sentiments: Institutional and Organizational Coevolution in the Early Thrift Industry." *American Journal of Sociology* 102:1606–51.

Heier, Siegfried. 2006. *Grid Integration of Wind Energy Convention Systems*. Chichester, U.K.: John Wiley.

Heinberg, Richard. 2005. *The Party's Over: Oil, War and the Fate of Industrial Societies*. Gabriola Island, B.C.: New Society Publishers.

Hess, David. 2007. *Alternative Pathways in Science and Industry: Activism, Innovation, and the Environment in an Era of Globalization*. Cambridge, Mass.: MIT Press.

Heymann, Matthias. 1998. "Signs of Hubris: The Shaping of Wind Technology Styles in Germany, Denmark, and the United States, 1940–1990." *Technology and Culture* 39:641–70.

Hirsh, Richard F. 1999. *Power Loss: The Origins of Deregulation and Restructuring in the American Electric Utility System*. Cambridge, Mass.: MIT Press.

Hoffman, Andrew J. 1997. *From Heresy to DogMass.: An Institutional History of Corporate Environmentalism*. San Francisco: New Lexington Press.

Hoffman, Andrew J., and Marc J. Ventresca. 2002. *Organizations, Policy, and the Natural Environment: Institutional and Strategic Perspectives*. Stanford, Calif.: Stanford University Press.

Hunt, Sally, and Graham Shuttleworth. 1996. *Competition and Choice in Electricity*. Chichester, U.K.: John Wiley.

Hunt, Scott, and Robert Benford. 2004. "Collective Identity, Solidarity and Commitment." In *The Blackwell Companion to Social Movements*, edited by David Snow, Sarah Soule, and Hanspeter Kriesi, 433–57. Malden, Mass.: Blackwell Publishing.

Jacobsson, Steffan, and Volkmar Lauber. 2005. "Germany: From a Modest Feed-in Law to a Framework for Transition." In *Switching to Renewable power: A Framework for the 21st Century*, edited by Volkmar Lauber, 122–59. London: Earthscan.

Jamison, Andrew. 2001. *The Making of Green Knowledge: Environmental Politics and Cultural Transformation*. Cambridge: Cambridge University Press.

Jamison, Andrew, Ron Eyerman, and Jacqueline Cramer, with Jeppe Læssøe. 1990. *The Making of the New Environmental Consciousness: A Comparative Study of the Environmental Movements in Sweden, Denmark, and the Netherlands*. Edinburgh: Edinburgh University Press.

Jiménez, Manuel. 1999. "Consolidation through Institutionalization? Dilemmas of the Spanish Environmental Movement in the 1990s." In *Environmental Movements: Local, National, and Global*, edited by Christopher Rootes, 149–71. London: Frank Cass.

———. 2003. "Spain." In *Environmental Protest in Western Europe*, edited by Christopher Rootes, 166–99. Oxford: Oxford University Press.

———. 2007. "The Environmental Movement in Spain: A Growing Force of Contention." *South European Society and Politics* 12 (3): 359–78.

Johnson, Anna, and Staffan Jacobsson. 2000. "The Emergence of a Growth Industry: A Comparative Analysis of the German, Dutch and Swedish Wind Turbine Industries." Department of Industrial Dynamics, Chalmers University of Technology, Göteborg.

Johnson, Erik, and John McCarthy. 2005. "The Sequencing of Transnational and National Social Movement Mobilization: The Organizational Mobilization of the Global and U.S. Environmental Movements." In

Transnational Protest and Global Activism, edited by Donatella della Porta and Sidney Tarrow, 71–93. Lanham, Md.: Rowman and Littlefield.

Joppke, Christian. 1993. *Mobilizing against Nuclear Energy: A Comparison of Germany and the United States*. Berkeley: University of California Press.

Karnøe, Peter. 1990. "Industrial Innovation and Incremental Learning: The Case of the Danish Wind Technology from 1975 to 1988." Paper presented at the Second International Conference on Management and Technology, Miami, Florida, February 27–March 2.

Karnøe, Peter, Peer Hull Kristensen, and Poul Houman Andersen, eds. 1999. *Mobilizing Resources and Generating Competencies: The Remarkable Success of Small and Medium-sized Enterprises in the Danish Business System*. Herndon, Va.: Copenhagen Business School Press, Books International.

Kealey Edward. 1987. *Harvesting the Air: Windmill Pioneers In Twelfth-Century England*. Berkeley: University of California Press.

Kendall, Bert, and Steven Nadis. 1980. *Energy Strategies: Toward a Solar Future, A Report of the Union of Concerned Scientists*. Cambridge, Mass.: Ballinger.

King, Brayden. 2008. "A Political Mediation Model of Corporate Response to Social Movement Activism." *Administrative Science Quarterly* 53:395–421.

King, Brayden, and Sarah Soule. 2007. "Social Movements as Extra-institutional Entrepreneurs: The Effect of Protests on Stock Price Returns." *Administrative Science Quarterly* 52:413–42.

Kirby, Brendan. 2007. "Evaluating Transmission Costs and Wind Benefits in Texas: Examining the ERCOT CREZ Transmission Study." http://www.consultkirby.com/files/Evaluating_Transmission_Costs_In_Texas.pdf (accessed February 2008).

Klandermans, Bert, Hanspeter Kriesi, and Sidney Tarrow. 1988. *From Structure to Action: Comparing Social Movement Research across Cultures*. Greenwich, Conn.: JAI Press.

Kolb, Felix. 2007. *Protest and Opportunities: The Political Outcomes of Social Movements*. Frankfurt: Campus Verlag.

Komor, Paul. 2006. "Wind Power in Colorado: Small Steps towards Sustainability." In *Sustainable Energy and the States: Essays on Politics, Markets, and Leadership*, edited by Dianne Rahm, 127–49. Jefferson, N.C.: McFarland and Company.

Kriesi, Hanspeter, Ruud Koopmans, Jan Willem Duyvendak, and Marco Giugni. 1995. *New Social Movements in Western Europe: A Comparative Analysis*. Minneapolis: University of Minnesota Press.

Lancaster, Thomas. 1989. *Policy Stability and Democratic Change: Energy in Spain's Transition*. University Park: Pennsylvania State University Press.

Lauber, Volkmar, ed. 2005. *Switching to Renewable Power: A Framework for the 21st Century*. London: Earthscan.

Lauber, Volkmar, and Lutz Mez. 2006. "Renewable Electricity Policy in Germany, 1974 to 2005." *Bulletin of Science, Technology and Society* 26 (2): 105–20.

Lévêque, Francois. 2006. *Competitive Electricity Markets and Sustainability.* Cheltenham, U.K.: Edward Elgar.

Levy, David, and Sandra Rothenberg. 2002. "Heterogeneity and Change in Environmental Strategy: Technological and Political Responses to Climate Change in the Global Automobile Industry." In *Organizations, Policy, and the Natural Environment: Institutional and Strategic Perspectives,* edited by Andrew J. Hoffman and Marc J. Ventresca, 173–94. Palo Alto, Calif.: Stanford University Press.

Locke, Richard. 1994. *Rebuilding the Economy: Local Politics and Industrial Change in Contemporary Italy.* Ithaca, N.Y.: Cornell University Press.

Lopez-Pintor, Rafael, and Luis Ramallo. 1986. "Public Acceptance of New Technologies in Spain." In *Public Acceptance of New Technologies: An International Review,* edited by Roger Williams and Stephen Mills, 353–63. London: Croom Helm.

Lounsbury, Michael. 2001. "Cultural Entrepreneurship: Stories, Legitimacy, and the Acquisition of Resources." *Strategic Management Journal* 22:545–64.

Lounsbury, Michael, Marc J. Ventresca, and Paul Hirsch. 2003. "Social Movement, Field Frames and Industry Emergence: A Cultural-Political Perspective on US Recycling." *Socio-Economic Review* 1:71–104.

Lovins, Amory B. 1977. *Soft Energy Paths: Toward a Durable Peace.* Cambridge, Mass.: Ballinger.

Lowe, Philip, and Jane Goyder. 1984. *Environmental Groups in Politics.* London: Allen and Unwin.

Lutzenhiser, Loren. 2001. "The Contours of U.S. Climate Non-policy." *Society and Natural Resources* 14:511–23.

MacKay, David J. C. 2008. *Sustainable Energy—Without the Hot Air.* http://www.withouthotair.com/.

Mainieri, Tina, Elaine Barnett, Trisha Valdero, John Unipan, and Stuart Oskamp. 1997. "Green Buying: The Influence of Environmental Concern on Consumer Behavior." *Journal of Social Psychology* 137:189–204.

Mason, Robert. 1998. "Whither Japan's Environmental Movement? An Assessment of Problems and Prospects at the National Level." *Pacific Affairs* 72:187–207.

Mayer, Rudd, Eric Blank, and Blair Swezey. 1999. "The Grassroots Are Greener: A Community-Based Approach to Marketing Green Power." Research report for the Renewable Energy Policy Project. http://www.repp.org/repp_pubs/pdf/grasgrnr.pdf (accessed April 2009).

McAdam Doug. 1988. *Freedom Summer: The Idealists Revisited.* Oxford: Oxford University Press.

McAdam, Doug, Sidney Tarrow, and Charles Tilly. 2001. *Dynamics of Contention.* Cambridge: Cambridge University Press.

McCammon, Holly, Karen Campbell, Ellen Granberg, and Christine Mowery. 2001. "How Movements Win: Gendered Opportunity Structures and U.S. Women's Suffrage Movements, 1866–1919." *American Sociological Review* 66:49–70.

McCright, Aaron M., and Riley E. Dunlap. 2000. "Challenging Global Warming as a Social Problem: An Analysis of the Conservative Movement's Counter-Claims." *Social Problems* 47 (4): 499–522.

————. 2003. "Defeating Kyoto: The Conservative Movement's Impact on U.S. Climate Change Policy." *Social Problems* 50 (3): 348–73.

Mehta, Michael. 2005. *Risky Business: Nuclear Power and Public Protest in Canada*. Lanham, Md.: Lexington Books.

Melluci, Alberto. 1989. *Nomads of the Present: Social Movements and Individual Needs in Contemporary Society*. London: Hutchinson Radius.

Mendonça, Miguel. 2007. *Feed-in Tariffs: Accelerating the Deployment of Renewable Energy*. London: Earthscan.

Meyer, David S., and Sidney Tarrow. 1998. "A Movement Society: Contentious Politics for a New Century." In *The Social Movement Society: Contentious Politics for a New Century*, edited by David S. Meyer and Sidney Tarrow. Lanham, Md.: Rowman and Littlefield.

Meyer, John W., David John Frank, Ann Hironaka, Evan Schofer, and Nancy Brandon Tuma. 1997. "The Structuring of a World Environmental Regime, 1870–1990." *International Organization* 51:623–51.

Meyer, John W., and Brian Rowan. 1977. "Institutionalized Organizations: Formal Structure as Myth and Ceremony." *American Journal of Sociology* 83:340–63.

Meyer, Niels I. 2004. "Renewable Energy Policy in Denmark." *Energy for Sustainable Development* 8:25–35.

Michaelowa, Axel. 2005. "The German Wind Energy Lobby: How to Promote Costly Technological Change Successfully." *European Environment* 15:192–99.

Minkoff, Debra. 1999. "Bending with the Wind: Strategic Change and Adaptation by Women's and Racial Minority Organizations." *American Journal of Sociology* 104:1666–1703.

Mitchell, Robert, Riley Dunlap, and Angela Mertig. 1992. "Twenty Years of Environmental Mobilization: Trends among National Environmental Organizations." In *American Environmentalism: The U.S. Environmental Movement, 1970–1990*, edited by Riley Dunlap and Angela Mertig, 11–26. Bristol, Pa.: Taylor and Francis.

Moore Kelly. 1999. "Political Protest and Institutional Change: The Anti-Vietnam War Movement and American Science." In *How Social Movements Matter*, edited by Marco Giugni, Doug McAdam, and Charles Tilly, 97–118. Minneapolis: University of Minnesota Press.

Murmann, Johann Peter. 2003. *Knowledge and Competitive Advantage: The Coevolution of Firms, Technology, and National Institutions*. Cambridge: Cambridge University Press.

Murtha, Thomas, Stephanie Lenway, and Jeffrey Hart. 2001. *Managing New Industry Creation: Global Knowledge Formation and Entrepreneurship in High Technology*. Palo Alto, Calif.: Stanford University Press.

Newell, Peter. 2000. *Climate for Change: Non-state Actors and the Global Politics of the Greenhouse*. Cambridge: Cambridge University Press.

Nielsen, Henrik Kaare. 2006. "Youth and the Antinuclear Power Movement in Denmark and West Germany." In *Between Marx and Coca-Cola: Youth Cultures in Changing European Societies, 1960–1980*, edited by Axel Schildt and Detlef Siegfried, 203–23. New York: Berghahn Books.

Nielsen, Kristian Hvidtfelt. 2005. "Danish Wind Power Policies from 1976 to 2004: A Survey of Policymaking and Techno-economic Innovation." In *Switching to Renewable power: A Framework for the 21st Century*, edited by Volkmar Lauber, 99–121. London: Earthscan.

Nitin, Nohria, and Robert G. Eccles. 1992. *Networks and Organizations: Structure, Form, and Action*. Boston: Harvard Business School Press.

Oliver, Pamela, and Daniel Myers. 2003. "Networks, Diffusion, and Cycles of Collective Action." In *Social Movements and Networks: Relational Approaches to Collective Action*, edited by Diani Mario and Doug McAdam, 173–203. New York: Oxford University Press.

Podobnik, Bruce. 2006. *Global Energy Shifts: Fostering Sustainability in a Turbulent Age*. Philadelphia: Temple University Press.

Poortinga, Wouter, Linda Steg, and Charles Vlek. 2004. "Values, Environmental Concern, and Environmental Behavior: A Study into Household Energy Use." *Environment and Behavior* 36:70–93.

Porter Michael. 1990. *The Competitive Advantage of Nations*. New York: Free Press.

Powell, Walter W. 1990. "Neither Market Nor Hierarchy: Network Forms of Organization." *Research in Organizational Behavior* 12:295–336.

———. 1991. "Expanding the Scope of Institutional Analysis." In *The New Institutionalism in Organizational Analysis*, edited by Walter W. Powell and Paul J. DiMaggio, 183–203. Chicago: University of Chicago Press.

Powell, Walter W., and Paul J. DiMaggio, eds. 1991. *The New Institutionalism in Organizational Analysis*. Chicago: University of Chicago Press.

Powers, Daniel A., and Yu Xie. 2000. *Statistical Methods for Categorical Data Analysis*. San Diego: Academic Press.

Preglau, Max. 1994. "The State and the Anti-nuclear Power Movement in Austria." In *States and Anti-nuclear Movements*, edited by Helena Flam, 37–69. Edinburgh: Edinburgh University Press.

Rader, Nancy, and Richard Norgaard. 1996. "Efficiency and Sustainability in Restructured Electric Markets: The Renewables Portfolio Standard." *Electricity Journal* (July).

Raeburn, Nicole C. 2004. *Changing Corporate America from Inside Out: Lesbian and Gay Workplace Rights*. Minneapolis: University of Minnesota Press.

Rao, Hayagreeva. 2009. *Market Rebels: How Activists Make or Break Radical Innovations*. Princeton, N.J.: Princeton University Press.

Rao, Hayagreeva, Philippe Monin, and Rodolphe Durand. 2003. "Institutional Change in Toque Ville: Nouvelle Cuisine as an Identity Movement in French Gastronomy." *American Journal of Sociology* 108:795–843.

Redlinger, Robert Y., Per Dannemand Andersen, and Poul Erik Morthorst. 2002. *Wind Energy in the 21st Century: Economics, Policy, Technology, and the Changing Electricity Industry*. New York: Palgrave.

Rickerson, Wilson H., Janet L. Sawin, and Robert C. Grace. 2007. "If the Shoe FITs: Using Feed-in Tariffs to Meet U.S. Renewable Electricity Targets." *Electricity Journal* 20 (4): 73–86.

Righter, Robert W. 1996. *Wind Energy in America: A History*. Norman: University of Oklahoma Press.

Roberts, Paul. 2005. *The End of Oil: On the Edge of a Perilous New World.* Boston: Houghton Mifflin.

Rochon, Thomas R. 1998. *Culture Moves: Ideas, Activism, and Changing Values.* Princeton, N.J.: Princeton University Press.

Rootes, Christopher. 2003. "Britain." In *Environmental Protest in Western Europe,* edited by Christopher Rootes, 20–58. Oxford: Oxford University Press.

———, ed. 2003b. *Environmental Protest in Western Europe.* Oxford: Oxford University Press.

———. 2003c. "The Transformation of Environmental Activism: An Introduction." In *Environmental Protest in Western Europe,* edited by Christopher Rootes, 1–20. Oxford: Oxford University Press.

———. 2004. "Environmental Movements." In *The Blackwell Companion to Social Movements,* edited by David Snow, Sarah Soule, and Hanspeter Kriesi, 608–40. Malden, Mass.: Blackwell Publishing.

Rosenbaum, Walter. 1998. *Environmental Politics and Policy.* Washington, D.C.: CQ Press.

Rowlands, Ian. 2005. "Global Climate Change and Renewable Energy: Exploring the Links." In *Switching to Renewable power: A Framework for the 21st Century,* edited by Volkmar Lauber, 62–82. London: Earthscan.

Roy, William G. 1997. *Socializing Capital: The Rise of the Large Industrial Corporation in America.* Princeton, N.J.: Princeton University Press.

Rucht, Dieter. 1994. "The Anti-nuclear Power Movement and the State in France." In *States and Anti-nuclear Movements,* edited by Helena Flam, 129–63. Edinburgh: Edinburgh University Press.

———. 1999. "The Impact of Environmental Movements in Western Societies." In *How Social Movements Matter,* edited by Marco Giugni, Doug McAdam, and Charles Tilly, 204–25. Minneapolis: University of Minnesota Press.

Rucht, Dieter, and Jochen Roose. 2003. "Germany." In *Environmental Protest in Western Europe,* edited by Christopher Rootes, 80–108. Oxford: Oxford University Press.

Rüdig, Wolfgang. 1990. *Anti-nuclear Movements: A World Survey of Opposition to Nuclear Energy.* Harlow, U.K.: Longman Current Affairs.

———. 1994. "Maintaining a Low Profile: The Anti-nuclear Movement and the British State." In *States and Anti-nuclear Movements,* edited by Helena Flam, 70–101. Edinburgh: Edinburgh University Press.

Russo, Michael. 2001. "Institutions, Exchange Relations, and the Emergence of New Fields: Regulatory Policies and Independent Power Production in America, 1978–1992." *Administrative Science Quarterly* 46 (2001): 57–86.

Ruttan, Vernon W. 2001. *Technology, Growth, and Development: An Induced Innovation Perspective.* New York: Oxford University Press.

Saxenian, AnnaLee. 2000. *Regional Advantage: Culture and Competition in Silicon Valley and Route 128.* Cambridge, Mass.: Harvard University Press (Ninth Edition.)

Schneiberg, Marc. 2002. "Organizational Heterogeneity and the Production of New Forms: Politics, Social Movements, and Mutual Companies in American

Fire Insurance, 1900–1930." In *Social Structure and Organizations Revisited: Research in the Sociology of Organizations* 19, edited by Michael Lounsbury and Marc J. Ventresca, 39–89. New York: Elsevier/JAI Press.

Schreurs, Miranda. 2003. *Environmental Politics in Japan, Germany, and the United States.* Cambridge: Cambridge University Press.

Scott, Richard. 2001. *Institutions and Organizations.* Thousand Oaks, Calif.: Sage.

———. 2002. *Organizations: Rational, Natural, and Open Systems,* 5th ed. Upper Saddle River, N.J.: Prentice Hall.

Scott, Richard, Martin Ruef, Peter Mendel, and Carol Caronna. 2000. *Institutional Change and Healthcare Organizations: From Professional Dominance to Managed Care.* Chicago: University of Chicago Press.

Scully, Maureen, and Amy Segal. 2002. "Passion with an Umbrella: Grassroots Activists in the Workplace." In *Social Structure and Organizations Revisited: Research in the Sociology of Organizations* 19, edited by Michael Lounsbury and Marc J. Ventresca, 125–68. New York: Elsevier/JAI Press.

Sine, Wesley, and Brandon Lee. 2009. "Tilting at Windmills? The Environmental Movement and the Emergence of the U.S. Wind Energy Sector." *Administrative Science Quarterly* 54:123–55.

Sioshansi, Fereidoon. 2008. *Competitive Electricity Markets: Design, Implementation, Performance.* Amsterdam: Elsevier.

Smeloff, Ed, and Peter Asmus. 1997. *Reinventing Electric Utilities: Competition, Citizen Action, and Clean Power.* Washington D.C.: Island Press.

Smith, Jackie. 2008. *Social Movements for Global Democracy.* Baltimore: Johns Hopkins University Press.

Smith, Jackie, and Dawn Wiest. 2005. "The Uneven Geography of Global Civil Society: National and Global Influences on Transnational Association." *Social Forces* 84:621–52.

Smith, Ted, David A. Sonnenfeld, and David Naguib Pellow. 2006. *Challenging the Chip: Labor Rights and Environmental Justice in the Global Electronics Industry.* Philadelphia: Temple University Press.

Snow, David, Burke Rochford, Steven Worden, and Robert Benford. 1986. "Frame Alignment Processes, Micromobilization, and Movement Participation." *American Sociological Review* 51:464–81.

Söder Lennart and Thomas Ackermann. 2005. "Wind Power in Power Systems: An Introduction." In *Wind Power in Power Systems,* edited by Thomas Ackermann, 25–51. John Wiley and Sons, Ltd.

Sorenson, Olav, and Pino G. Audia. 2000. "The Social Structure of Entrepreneurial Activity: Geographic Concentration of Footwear Production in the United States, 1940–1989." *American Journal of Sociology* 106:424–62.

Soule Sarah. 1997. "The Student Divestment Movement in the United States and Tactical Diffusion: The Shantytown Protest." *Social Forces,* 75: 855–883.

Soule, Sarah. 2009. *Contention and Corporate Social Responsibility.* New York: Cambridge University Press.

Soule, Sarah, Doug McAdam, John McCarthy, and Yang Su. 1999. "Protest Events: Cause or Consequence of the U.S. Women's Movement and Federal Congressional Activities, 1956–1979." *Mobilization* 4:239–56.

Soule, Sarah, and Susan Olzak. 2004. "When Do Movements Matter? The Politics of Contingency and the Equal Rights Amendment." *American Sociological Review* 69:473–98.

Sovacool, Benjamin. 2008. "Valuing the Greenhouse Gas Emissions from Nuclear Power: A Critical Survey." *Energy Policy* 36:2950–63.

Sovacool, Benjamin, Hans Lindboe, and Ole Odgaard. 2008. "Is the Danish Wind Energy Model Replicable for Other Countries?" *Electricity Journal* 21:27–38.

Spencer, Jennifer, Thomas Murtha, and Stephanie Lenway. 2005. "How Governments Matter to New Industry Creation." *Academy of Management Review* 30:321–37.

Stets, Jan, and Chris Biga. 2003. "Bringing Identity Theory into Environmental Sociology." *Sociological Theory* 21:398–423.

Storchmann, Karl. 2005. "The Rise and Fall of German Hard Coal Subsidies." *Energy Policy* 33 (11): 1469–92.

Strang David and Dong-Il Jung. 2005. "Organizational Change as an Orchestrated Social Movement: Recruitment to a Corporate Quality Initiative." In *Social Movements and Organization Theory*, edited by Gerald Davis, Doug McAdam, Richard Scott, and Mayer Zald, 280–309. Cambridge: Cambridge University Press.

Sutton, John, and Frank Dobbin. 1996. "The Two Faces of Governance: Responses to Legal Uncertainty in American Firms, 1955–1985." *American Sociological Review* 61:794–811.

Tamplin, Arthur, and Thomas Cochran. 1974. *Radiation Standards of Hot Particles*. Washington, D.C.: Natural Resources Defense Council.

Tarrow, Sidney. 1994. *Power in Movement: Social Movements and Contentious Politics*. Cambridge: Cambridge University Press.

Tertzakian, Peter. 2006. *A Thousand Barrels a Second: The Coming Oil Break Point and the Challenges Facing an Energy Dependent World*. New York: McGraw-Hill.

Thornton, Patricia, and William Ocasio. 1999. "Institutional Logics and the Historical Contingency of Power in Organizations: Executive Succession in the Higher Education Publishing Industry, 1958–1990." *American Journal of Sociology* 105:801–43.

Tilly, Charles, and Sidney Tarrow. 2007. *Contentious Politics*. Boulder, Colo.: Paradigm.

Toke, David, and Volkmar Lauber. 2007. "Anglo-Saxon and German Approaches to Neoliberalism and Environmental Policy: The Case of Financing Renewable Energy." *Geoforum* 38:677–87.

Van der Heijden, Hein-Anton. 1994. "The Dutch Nuclear Energy Conflict 1973–1989." In *States and Anti-nuclear Movements*, edited by Helena Flam, 101–28. Edinburgh: Edinburgh University Press.

———. 1999. "Environmental Movements, Ecological Modernization and Political Opportunity Structures." *Environmental Politics* 8:199–221.

Van de Ven, Andrew, and Raghu Garud. 1989. "A Framework for Understanding the Emergence of New Industries." In *Research on Technological Innovation, Management and Policy*, 4:195–225. Greenwich, Conn.: JAI Press.

Vasi, Ion Bogdan. 2006. "Organizational Environments, Framing Processes, and the Diffusion of the Program to Address Global Climate Change among

Local Governments in the United States." *Sociological Forum* 21 (3): 439–66.

———. 2007. "Thinking Globally, Planning Nationally, and Acting Locally: Nested Organizational Fields and the Adoption of Environmental Practices." *Social Forces* 86 (1): 113–37.

———. 2009. "Social Movements and Industry Development: The Environmental Movement's Impact on the Wind Energy Industry." *Mobilization: The International Quarterly Review of Social Movement Research* 14(3): 315–36.

Wagner, Peter. 1994. "Contesting Policies and Redefining the State: Energy Policy-Making and the Anti-nuclear Movement in West Germany." In *States and Anti-nuclear Movements*, edited by Helena Flam, 264–95. Edinburgh: Edinburgh University Press.

White, Harrison. 1981. "Where Do Markets Come From?" *American Journal of Sociology* 87:517–47.

———. 2004. *Markets from Networks: Socioeconomic Models of Production.* Princeton, N.J.: Princeton University Press.

Wikle, Thomas. 1995. "Geographical Patterns of Membership in U.S. Environmental Organizations." *Professional Geographer* 47 (1): 41–48.

Williams Wendy and Robert Whitcomb. 2007. *Cape Wind, Money, Celebrity, Class, Politics, and the Battle for Our Energy Future on Nantucket Sound.* New York: Public Affairs.

Wilson, Jeremy. 2002. "Continuity and Change in the Canadian Environmental Movement: Assessing the Effects of Institutionalization." In *Canadian Environmental Policy: Contexts and Cases*, edited by Deborah VanNijnatten and Robert Boardman, 2nd ed., 46–65. Toronto: Oxford University Press.

Wiser, Ryan, Christopher Namovicz, Mark Gielecki, and Robert Smith. 2007. "Renewables Portfolio Standards: A Factual Introduction to Experience from the United States." Ernest Orlando Lawrence Berkeley National Laboratory. http://eetd.lbl.gov/ea/emp/reports/62569.pdf (accessed May 2010).

Wiser, Ryan, Kevin Porter, and Steve Clemmer. 2000. "Emerging Markets for Renewable Energy: The Role of State Policies during Restructuring." *Electricity Journal* (January/February 2000).

Zaccour, Georges. 1998. *Deregulation of Electric Utilities.* Boston: Kluwer Academic.

Zald, Mayer, Calvin Morrill, and Hayagreeva Rao. 2005. "The Impact of Social Movements on Organizations: Environment and Responses." In *Social Movements and Organization Theory*, edited by Gerald Davis, Doug McAdam, Richard Scott, and Mayer Zald, 253–79. New York: Cambridge University Press.

Index